T0279350

CLOUD MANAGEMENT AND SECURITY

CLOUD MANAGEMENT AND SECURITY

Imad M. Abbadi

University of Oxford, UK

WILEY

This edition first published 2014
© 2014 John Wiley & Sons, Ltd

Registered office
John Wiley & Sons Ltd, The Atrium, Southern Gate, Chichester, West Sussex, PO19 8SQ, United Kingdom

For details of our global editorial offices, for customer services and for information about how to apply for permission to reuse the copyright material in this book please see our website at www.wiley.com.

The right of the author to be identified as the author of this work has been asserted in accordance with the Copyright, Designs and Patents Act 1988.

All rights reserved. No part of this publication may be reproduced, stored in a retrieval system, or transmitted, in any form or by any means, electronic, mechanical, photocopying, recording or otherwise, except as permitted by the UK Copyright, Designs and Patents Act 1988, without the prior permission of the publisher.

Wiley also publishes its books in a variety of electronic formats. Some content that appears in print may not be available in electronic books.

Designations used by companies to distinguish their products are often claimed as trademarks. All brand names and product names used in this book are trade names, service marks, trademarks or registered trademarks of their respective owners. The publisher is not associated with any product or vendor mentioned in this book.

Limit of Liability/Disclaimer of Warranty: While the publisher and author have used their best efforts in preparing this book, they make no representations or warranties with respect to the accuracy or completeness of the contents of this book and specifically disclaim any implied warranties of merchantability or fitness for a particular purpose. It is sold on the understanding that the publisher is not engaged in rendering professional services and neither the publisher nor the author shall be liable for damages arising herefrom. If professional advice or other expert assistance is required, the services of a competent professional should be sought.

Library of Congress Cataloging-in-Publication Data

Abbadi, Imad M., author.
  Cloud management and security / Imad M. Abbadi.
    pages cm
  Includes bibliographical references and index.
  ISBN 978-1-118-81709-4 (hardback)
  1. Cloud computing. 2. Web services. 3. Computer security.   I. Title.
  QA76.585.A23 2014
  004.67′82–dc23

                                        2014011409

A catalogue record for this book is available from the British Library.

ISBN: 9781118817094

Set in 10/12pt Times by Aptara Inc., New Delhi, India

1   2014

Contents

Part Three   PRACTICAL EXAMPLES

About the Author

Dr. Imad Abbadi is an Associate Professor of Information Security with more than 18 years' experience of leading enterprise-scale projects. He works at Oxford University, leading activities to establish the next-generation trustworthy Cloud infrastructure. He has pioneered a novel, worldwide course in Cloud security which has been adopted at the university.

Dr. Abbadi currently teaches his Cloud security course as part of Oxford University's M.Sc. in Software and Systems Security. In addition to his teaching role he is also a principal consultant and senior project manager for enterprise-scale projects spanning several domains, such as finance and healthcare. Dr. Abbadi is a strategic planner who helps several organizations to define their Cloud adoption strategy. Further, he has invented several tools to enhance Cloud trustworthiness and authored more than 40 scientific papers.

Preface

Cloud computing is a new concept, building on well-established industrial technologies. The interactions between the technologies behind Cloud computing had never been of great interest in the academic domain before the Cloud era. The emergence of Cloud computing as an Internet-scale critical infrastructure has greatly encouraged the collaboration between industry and academia to analyze this infrastructure. Such collaborations would help in understanding the vulnerabilities of Cloud and defining research agendas to address the identified vulnerabilities. In fact, funding bodies and governments have already allocated generous grants to encourage both academic and industrial collaboration on research activities in Cloud computing. In addition, some universities have very recently introduced Cloud computing-related subjects as part of their undergraduate and postgraduate degrees to advance the knowledge in this domain.

Cloud computing has emerged from industry to academia without transferring the knowledge behind this domain. This results in confusion and misunderstanding. Most of the available trusted resources are industrial and scattered around hundreds of technical manuals and white papers. These cover different complex domains (e.g., infrastructure management, distributed database management systems, clustering technology, software architecture, security management, and network management). These domains are not easy to understand, as integrated science, for many people working both in the industry and academia. This book does not discuss the complex details of each technical element behind Cloud computing, as these are too complicated to be covered in a single textbook. In addition, discussing these will not help non-technical readers to understand Cloud computing. This book rather provides a conceptual and integrated view of the overall Cloud infrastructure; it covers Clouds structure, operation management, property and security. It also discusses trust in the Cloud – that is, how to establish trust in Clouds using current technologies – and presents a set of integrated frameworks for establishing next-generation trustworthy Cloud computing. These elements have never been discussed before in the same way. The book is rich in real-life scenarios, currently used in a Cloud production environment. Moreover, we provide practical examples partly clarifying the concepts discussed throughout the book.

The main objective of this book is to establish the foundations of Cloud computing, building on an in-depth and diverse understanding of the technologies behind Cloud computing. The author has more than 15 years of senior industrial experience managing and building all technologies behind Cloud computing. The book is also based on strong scientific publication records at international conferences and in leading journals [1–17]. That is to say, this book presents a neutral view of the area, supported by solid scientific foundations and a strong industrial vision. Oxford University has adopted this book as part of its MSc in Software and Systems Security.

Guide to Using this Book

This section discusses the organization of the book and the required background when reading different chapters of the book. It also aims to help instructors seeking to adopt this book for their undergraduate or postgraduate course levels.

Organization of the Book

This book starts with an introduction, followed by three parts: Cloud management; Cloud security; and practical examples.

> The introduction is presented in **Chapter 1**. It discusses the fundamental concepts of Cloud computing. That is, Cloud definition, Cloud services, Cloud deployment types, and the main challenges in Clouds.

> The first part (i.e., Cloud management) consists of four chapters. **Chapter 2** presents the main components of the Cloud infrastructure. It also discusses the relationship between the components of Cloud and their interactions. This chapter is key to understanding the properties of Cloud, the real challenges of Cloud, and the differences between different deployment types of Cloud. **Chapter 3** analyzes Cloud's management platforms. The chapter starts by identifying and discussing the main services which are required to automatically manage Cloud resources. It then presents a unified view of Cloud's management platforms and discusses their required inputs. Following that, the chapter presents the process workflow of managing user requirements and identifying weaknesses in the management process. **Chapter 4** identifies and analyzes the main properties of the Cloud infrastructure. Such properties are important for Cloud users when comparing different Cloud providers. They are also important for Cloud providers when assessing their infrastructure and introducing various Cloud business models. Moreover, realizing the Cloud properties is very important when conducting research in the Cloud computing domain. Finally, **Chapter 5** discusses Clouds automated management services: virtual and application resource management services.

> The second part (i.e., Cloud security) consists of six chapters. **Chapter 6** introduces Part Two and highlights its relation to Part One of the book. It also briefly outlines the trusted computing principles. **Chapter 7** discusses the problem of establishing trustworthy Cloud. The chapter concludes with a set of research directions for establishing trust in Cloud. The remaining chapters in this part extend the identified directions and draw a set of integrated frameworks for establishing next-generation trustworthy Cloud computing. **Chapter 8** lays a foundation framework to address the question of how users can establish trust in Cloud without the need to get involved in complex technical details. **Chapter 9** discusses mechanisms for remote attestation in Cloud and addresses the question of how to establish trust in

a composition of multiple entities in which the entities could change dynamically. **Chapter 10** presents a framework for establishing a trustworthy provenance system. This helps in monitoring, verifying, and tracking the operation management of the Cloud infrastructure, that is it helps in the direction of proactive service management, finding the cause of incidents, customer billing assurance, security monitoring (as in the case of lessening the effects of insider threats), security and incident reporting, and tracking both management data and customer data across the infrastructural resources. **Chapter 11** discusses the problem of insiders; it provides a systematic method to identify potential and malicious insiders in a Cloud environment.

The last part (i.e., practical examples) consists of two chapters. **Chapter 12** presents real-life commercial and open-source examples of some of the concepts discussed in this book. It also presents a possible implementation of some of the concepts in the book. **Chapter 13** presents a case study which helps in understanding the concepts discussed throughout the book.

Required Background

Readers of this book should have a basic understanding of computer security principles and some understanding of computer systems architecture and network connectivity. Each chapter in Part Two is composed of three main components: problem analysis, a framework, and implementation protocols. The first two components require careful understanding of Part One, while the third component requires, in addition, an extensive understanding of trusted computing principles and cryptographic protocols. We introduce trusted computing principles in Chapter 6.

Suggestions for Course Organization

The layout of this book has been carefully designed for postgraduate studies. Specifically, most chapters cover the teaching material of the Cloud security module[1] of Oxford University's MSc in Software and Systems Security. This degree is specifically designed to fit the needs of industrial professionals. The book could also be of great benefit for undergraduate studies. We suggest the following layout in both cases.

Postgraduate Study

This could follow the Oxford University curriculum in teaching the book (available on its website), in which we cover selected parts from all chapters. Alternatively, the book could be taught as two modules: Cloud management and advanced Cloud security. The Cloud management module would need to complete the first part of the book and part of the third part of the book. The advanced Cloud security module would need to start with a high-level

---

[1] http://www.cs.ox.ac.uk/softeng/subjects/CLS.html (accessed March 2013).

introduction to the first part of the book and then cover the details of the second and third parts of the book. In addition, it would need to cover federated identity management and key management in Cloud and federated Clouds, which we do not cover here.

Undergraduate Study

As in the case of postgraduate studies, an undergraduate course could cover the Cloud computing subject in two modules: Cloud management and Cloud security. Cloud management could cover selected sections from all chapters of the first part of the book. The Cloud security module would assume that students had already studied information security and Cloud management. Cloud security could cover the problem analysis and framework components of the second part of the book. Undergraduate students would also benefit from the third part of the book as laboratory-based exercise work.

References

[1] Imad M. Abbadi. Middleware services at cloud application layer. In *IWTMP2PS '11: Proceedings of Second International Workshop on Trust Management in P2P Systems*. Kochi, India, July 2011.

[2] Imad M. Abbadi. Clouds infrastructure taxonomy, properties, and management services. In Ajith Abraham, Jaime Lloret Mauri, John F. Buford, Junichi Suzuki, and Sabu M. Thampi (eds), *Advances in Computing and Communications*, vol. 193 of *Communications in Computer and Information Science*, pp. 406–420. Springer-Verlag: Berlin, 2011.

[3] Imad M. Abbadi. Middleware services at cloud virtual layer. In *DSOC 2011: Proceedings of the 2nd International Workshop on Dependable Service-Oriented and Cloud Computing*. IEEE Computer Society, August 2011.

[4] Imad M. Abbadi. Operational trust in clouds' environment. In *MoCS 2011: Proceedings of the Workshop on Management of Cloud Systems*. IEEE, June 2011.

[5] Imad M. Abbadi. Self-Managed services conceptual model in trustworthy clouds' infrastructure. In *Workshop on Cryptography and Security in Clouds*. IBM, Zurich, March 2011. http://www.zurich.ibm.com/cca/csc2011/program.html.

[6] Imad M. Abbadi. Toward trustworthy clouds' internet scale critical infrastructure. In *ISPEC '11: Proceedings of the 7th Information Security Practice and Experience Conference*, vol. 6672 of *LNCS*, pp. 73–84. Springer-Verlag: Berlin, 2011.

[7] Imad M. Abbadi, Muntaha Alawneh, and Andrew Martin. Secure virtual layer management in clouds. In *The 10th IEEE International Conference on Trust, Security and Privacy in Computing and Communications (IEEE TrustCom-10)*, pp. 99–110. IEEE, November 2011.

[8] Imad M. Abbadi, Mina Deng, Marco Nalin, Andrew Martin, Milan Petkovic, Ilaria Baroni, and Alberto Sanna. Trustworthy middleware services in the cloud. In *CloudDB'11*. ACM Press: New York, 2011.

[9] Imad M. Abbadi and John Lyle. Challenges for provenance in cloud computing. In *3rd USENIX Workshop on the Theory and Practice of Provenance (TaPP '11)*. USENIX Association, 2011.

[10] Imad M. Abbadi and Andrew Martin. Trust in the cloud. *Information Security Technical Report*, 16(3–4):108–114, 2011.

[11] Imad M. Abbadi and Cornelius Namiluko. Dynamics of trust in clouds – challenges and research agenda. In *6th International Conference for Internet Technology and Secured Transactions (ICITST-2011)*, pp. 110–115. IEEE, December 2011.

[12] Imad M. Abbadi, Cornelius Namiluko, and Andrew Martin. Insiders analysis in cloud computing focusing on home healthcare system. In *6th International Conference for Internet Technology and Secured Transactions (ICITST-2011)*, pp. 350–357. IEEE, December 2011.

[13] Muntaha Alawneh and Imad M. Abbadi. Defining and analyzing insiders and their threats in organizations. In *2011 IEEE International Workshop on Security and Privacy in Internet of Things (IEEE SPIoT 2011)*. IEEE, November 2011.

[14] Imad M. Abbadi. Clouds trust anchors. In *11th IEEE International Conference on Trust, Security and Privacy in Computing and Communications (IEEE TrustCom-11)*. IEEE, June 2012.

[15] Imad M. Abbadi. A framework for establishing trust in cloud provenance. *International Journal of Information Security*, 11:1–18, 2012.

[16] Imad M. Abbadi and Muntaha Alawneh. A framework for establishing trust in the cloud. *Computers and Electrical Engineering Journal*, 38:1073–1087, 2012.

[17] Imad M. Abbadi and Anbang Ruan. Towards trustworthy resource scheduling in clouds. *Transactions on Information Forensics & Security*, in press.

Acknowledgments

The author would like to thank Andrew Martin for taking the initiative and introducing the Cloud security module within the University of Oxford, as part of the Department of Computer Science part-time MSc in Systems and Software Engineering. Andrew was the source of encouragement to complete this book, which is designed specifically to support this program of study.

Acronyms

ACaaS	Access Control as a Service
ADaaAS	Adaptability as an Application Service
ADaaVS	Adaptability as a Virtual Service
AIK	Attestation Identity Key
AVaaAS	Availability as an Application Service
AVaaVS	Availability as a Virtual Service
CCA	Cloud Client Agent
CCoT	Collaborating Domain Chain of Trust
CMD	Cloud Collaborating Management Domain
COD	Organization Collaborating Outsourced Domain
CoT	Chain of Trust
CRTM	Core Root of Trust for Measurement
CSA	Cloud Server Agent
DBMS	Database Management System
DC-C	Domain Controller Client Side
DCoT	Domain Chain of Trust
DC-S	Domain Controller Server Side
DR	Disaster Recovery
HD	Organization Home Domain
IaaS	Infrastructure as a Service
IR	Integrity Report
LaaS	Log as a Service
LaaSD	Log as a Service Domain
LCA	LaaS Client Agent
LSA	LaaS Server Agent
MD	Management Domain
MTT-Deploy	Mean Time to Deploy
MTBF	Mean Time Between Failure
MTTD	Mean Time to Discover
MTTF	Mean Time to Failure
MTTI	Mean Time to Invoke
MTTPHW	Mean Time to Procure Hardware Resources
MTTR	Mean Time to Recover
MTTS-Down	Mean Time to Scale Down

MTTS-UP	Mean Time to Scale Up
NAS	Network Attached Storage
NIST	National Institute of Standards & Technology
OD	Organization Outsourced Domain
OS	Operating System
PaaS	Platform as a Service
PTS	Platform Trust Service
PCR	Platform Configuration Register
PKL	Public Key List
RAC	Real Application Cluster
RBAC	Role-Based Access Control
RCoT	Resource Chain of Trust
RLaaAS	Reliability as an Application Service
RLaaVS	Reliability as a Virtual Service
RSaaAS	Resilience as an Application Service
RSaaVS	Resilience as a Virtual Service
SAaaVS	System Architect as a Virtual Service
SaaS	Software as a Service
SAN	Storage Area Network
SCaaAS	Scalability as an Application Service
SCaaVS	Scalability as a Virtual Service
SLA	Service Level Agreement
TCB	Trusted Computing Base
TCG	Trusted Computing Group
TCS	Trusted Computing Services
TCSD	Trusted Core Service Daemon
TP	Trusted Platform
TPM	Trusted Platform Module
VCC	Virtual Control Center
VM	Virtual Machine
VMA	Virtual Machine Agent
VMI	Virtual Machine Image
VMM	Virtual Machine Manager
vTPM	Virtual TPM

1

Introduction

This chapter introduces Cloud computing. The introduction helps the reader to get an overview of Cloud computing and its main challenges. Subsequent chapters of this book assume the reader understands the content of this chapter.

1.1 Overview

Cloud computing originates from industry (commercial requirements and needs). Governments and leading industrial bodies involved academia at early stages of adopting Cloud computing because of its promising future as an Internet-scale critical infrastructure. Involving academia would ensure that Cloud computing is critically analyzed, which helps in understanding its problems and limitations. This would also help in advancing the knowledge of this domain by defining and executing research road maps to establish next-generation trustworthy Cloud infrastructure. Moreover, academia would provide the required education in Cloud computing by developing undergraduate and postgraduate courses in this domain.

Cloud comes with enormous advantages; for example, it reduces the capital costs of newly established businesses, it reduces provisioning time of different types of services, it establishes new business models, it reduces the overhead of infrastructure management, and it extends IT infrastructures to the limits of their hosting Cloud infrastructure. Although Cloud computing is associated with such great features, it also has critical problems preventing its wider adoption by critical business applications, critical infrastructures, or even end-users with sensitive data. Examples of such problems include: security and privacy problems, operational management problems, and legal concerns. The immaturity of Cloud and the generosity of its allocated funds have made Cloud computing, in a relatively short period of time, one of the most in-demand research topics around the world.

Cloud computing is built on complex technologies which are not easy to understand, as an integrated science, for many people working in the industry and academia. A fundamental reason behind this is the lack of resources analyzing current Cloud infrastructure, its properties and limitations [1, 2]. The main objective of this book is to establish the foundations of Cloud computing, which would help researchers and professionals to understand Cloud as an

Cloud Management and Security, First Edition. Imad M. Abbadi.
© 2014 John Wiley & Sons, Ltd. Published 2014 by John Wiley & Sons, Ltd.
Companion Website: www.wiley.com/go/abbadi_cloud

integrated science. Understanding the Cloud structure and properties is key for conducting practical research in this area that could possibly be adopted by industry.

Most current research assumes Cloud computing is a black-box that has physical and virtual resources. The lack of careful understanding of the properties, structure, management, and operation of the black-box results in confusion and misunderstanding. In terms of misunderstanding, this relates to Cloud's limitations and the expectations of what it could practically provide. For example, some people claim that Cloud has immediate and unlimited capabilities, that is immediate and unlimited scalability. This is not practical considering present-day technologies, such as the limitations of hardware resources. There are also many other factors that have not been considered in such strong claims, for example should Cloud provide unlimited resources in case of application software bugs? Should resources be available immediately upon request without users' prior agreement? This book discusses these issues in detail.

This chapter is organized as follows. Section 1.2 discusses the definition of Cloud computing. Section 1.3 clarifies the evolution of Cloud computing. Section 1.4 discusses Cloud services. Section 1.5 discusses Cloud deployment types. Section 1.6 discusses the main challenges of Clouds. Finally, we summarize the chapter in Section 1.7 and provide a list of exercises in Section 1.8.

1.2   Cloud Definition

Cloud computing is a new buzzword in computing terms and it is associated with various definitions. In this book we focus on two definitions: the first is provided by the National Institute of Standards & Technology (NIST) [2] and the second is provided by an EU study of the future directions of Clouds [3]. The main reasons for analyzing these definitions in particular are:

- The good reputation of the organizations behind the definitions. For example, the EU study was edited by representatives of leading universities and industrial bodies such as Oracle, Google, Microsoft, and IBM.
- We found thsse definitions to be unique, such that their combination provides the most important elements of Cloud as covered throughout this book.

NIST defines Cloud as a model for enabling ubiquitous, convenient, on-demand network access to a shared pool of configurable computing resources (e.g., networks, servers, storage, applications, and services) that can be rapidly provisioned and released with minimal management effort or service provider interaction [2].

In contrast:

An EU study defines Cloud as an elastic execution environment of resources involving multiple stakeholders and providing a metered service and multiple granularities for specified level of quality [3].

Although both definitions come from reputable organizations, they are not consistent. This is not to say that either of them is wrong, but they are incomplete. Both definitions reveal many important keywords reflecting Clouds capabilities; however, a careful analysis of these definitions shows they only have one keyword in common. The first definition uses *'rapidly provisioned and released'* while the second definition uses *'elastic execution.'* These two keywords have the same objective. However, other keywords are not the same, for example 'minimal management effort' as stated by the NIST definition is not stated anywhere in the EU definition. Similarly, the EU definition uses the keyword 'metered service' which is again not stated anywhere in the NIST definition.

Cloud computing is in fact a combination of both definitions as each definition provides a partial view of the Cloud attributes. Therefore, we could redefine Cloud computing as follows.

Cloud computing is a model involving multiple stakeholders and enabling ubiquitous, convenient, on-demand network access to a shared pool of configurable computing resources (e.g., networks, servers, storage, applications, and services) that can be rapidly provisioned and released with minimal management effort or service provider interaction. The model provides a metered service and multiple granularities for a specified level of quality.

This book focuses primarily on the details behind the elements in the definition which would clarify the Cloud computing black-box.

1.3 Cloud Evolution

Enterprise infrastructures witnessed three major fundamental changes, which were a result of major innovations in computer science. These are as follows:

- *Traditional enterprise infrastructure.* This is the foundation of the virtualization era. Initially, it starts with a few powerful servers (what used to be called mainframes). With advances in technologies and an increased number of required applications, the number of servers increases rapidly. This results in a huge number of resources within an enterprise infrastructure. Despite the complexity of the traditional enterprise infrastructure, the relationship between customers and their resources is simple. Within this, the requirements of customers are carefully analyzed by system analysts. The system analysts forward the analyzed results to enterprise architects. The enterprise architects deliver an architecture which is designed to address the needs of a specific customer application requirement. The resources required by the delivered architecture in most cases run a specific customer applications. This process results in a one-to-one relationship between architecture and customer. Such a relationship causes huge wastage of resources including, for example, computational resources, power consumption, and data-center spaces. In contrast, this relationship results in a relatively more secure and customized design than the other evolution models of enterprise infrastructure.
- *Virtual enterprise infrastructure.* This is the foundation of today's Cloud infrastructure. The problems of the traditional enterprise infrastructure, which affect the green agenda, require novel innovations enabling customers to share resources without losing control or

increasing security risks. This was the start of the virtualization era, which brings tremendous advantages in terms of consolidating resources and results in effective utilization of power, data-center space, etc. A virtual enterprise infrastructure suffers from many problems, such as security, privacy, and performance problems, which restricts many applications from running on virtual machines. As a result, virtual infrastructures for many enterprises support applications that run on virtual resources and those that run directly on physical resources.

The virtualization era changes the mentality of enterprise architects as the relationship between users and their physical resources is no longer one-to-one. This raises a big challenge in terms of how such a consolidated virtualized architecture could satisfy users' dynamic requirements and unique application nature. Enterprise architects address this by studying the environment inherited from the traditional enterprise infrastructure, to find that different architectures have some similarities. The similarities between independent applications enable enterprise architects to split the infrastructure into groups. Each group has architecture-specific static properties. The properties enable the group to address common requirements of a certain category of applications. For example, a group could be allocated to applications that tolerate a single point of failure; another group could be allocated to applications that require full resilience with no single point of failure; a third group could be allocated to applications that are highly computational; a group for archiving systems; and so on.

The second part of the challenging question is how such a grouping, which is associated with almost static properties, could be used to address users' dynamic requirements and their unique application nature. Enterprise architects realize that virtualization can be fine-tuned and architected to support the dynamic application requirements which cannot be provided by the physical group static properties. In other words, a combination of static physical properties and dynamic virtual properties is used to support customer expectations in a virtual enterprise infrastructure.

- *Cloud infrastructure*. This has evolved from the virtual enterprise infrastructure. Chapters 2 and 4 cover the details of Cloud structure and its attributes. Clouds come with many important and promising features, such as direct interaction with customers via supplied APIs, automatically managed resources via self-managed services, and support for a pay-per-use model. In addition, Cloud computing comes with new promising business models that would enable more efficient utilization of resources and quicker time-to-market. Cloud computing inherits the problems of the virtual infrastructure and in addition, it comes with more serious problems including security problems, operational and data management problems. The problems associated with Cloud prevent its wider adoption, especially by critical organizations. This chapter discusses the most important problems in Clouds.

1.4   Cloud Services

Cloud services are also referred to as Cloud types in some references. These are served by Cloud providers to their customers following a pre-agreed service level agreement (SLA). Figure 1.1 illustrates the commonly agreed Cloud services in the context of a Cloud environment. Understanding these services requires understanding the structure of the Cloud, which is discussed in detail in Chapter 2. As illustrated in the figure, the Cloud structure could be viewed

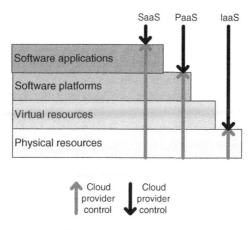

Figure 1.1 Cloud services

based on the hosting relationship as the following hierarchical layers: physical layer, virtual layer, software platform and software application layer. The physical layer is composed of all physical components and their management software components, including the operating system and the hypervisor. The virtual layer is composed of virtual machines, virtual storage, and a virtual network managed by the physical layer. The software application and software platforms are self-explanatory, and could be served either by the virtual layer or directly by the physical layer.

The management of Cloud services is a shared responsibility between the Cloud provider and their customers. The level of responsibility is Cloud service specific, as explained next. Cloud computing has the following main services.

- *Infrastructure as a service (IaaS)*. IaaS provides virtual compute and store resources as a service to customers. Cloud providers in IaaS manage the physical resources and their hypervisors. Cloud customers run their software stack and manage the content of their allocated virtual resources, including guest operating system. Customers in this type should, in principle, have overall control of their data. At the time of writing, Cloud providers have ultimate control of customer data.
- *Platform as a service (PaaS)*. PaaS provides the environment and software platforms that Cloud customers can use to develop and host their own software applications. Unlike IaaS, PaaS customers do not manage the software platforms provided by the Cloud, but only need to manage their own software stack. Cloud providers of PaaS expose their own APIs, which are used directly by customer applications. The exposed APIs, at the time of writing, do not follow any standard. As a result, Cloud customers of PaaS cannot move their applications transparently across competing Cloud providers.
- *Software as a service (SaaS)*. SaaS provides ready-to-consume software applications which address the needs of specific business functions and processes. Cloud providers manage the software applications and the hosting environment completely. Cloud customers might need to manage their specific configurations within the supported software application.

We conclude from the above that Cloud computing provides full outsourcing support for SaaS, partial outsourcing support for PaaS, and minimal outsourcing support for IaaS. That is, IaaS in theory provides customers with the greatest control over their resources, while SaaS provides Cloud providers with the greatest control over their customers' data and Cloud customers with the least control over their resources.

The above services are the main services a Cloud provider supports. Some references discuss other services, such as backup as a service, log as a service, etc. These services would be categorized under the above main services. For example, backup as a service could be viewed as SaaS. A Cloud customer does not necessarily need to stick to one service. It is, rather, likely for a Cloud customer to have a combination of different services. The selection of the service should be based on different complex factors, such as: the nature of the hosted application that will be using the service, the customer level of competence in using IT, the desired level of control, security and privacy requirements, cost factors, and legal requirements. We discuss these in detail throughout the book.

1.5   Cloud Deployment Types

Clouds have the following main deployment types (also referred to as deployment models in some references):

- *Public Cloud.* The infrastructure of a public Cloud is owned by the Cloud provider, but leased to Cloud customers. The Cloud provider typically manages its physical infrastructure, but it could outsource specific functions to a third party as in the case of outsourcing hardware maintenance. Example of this type includes Amazon and RackSpace.
- *Private Cloud.* A private Cloud deployment type is owned and used by a specific enterprise. That is, the enterprise employees are the only customer of the private Cloud. The private Cloud could either directly manage its own infrastructure or it could outsource the management to a third party. Example of a private Cloud deployment type includes most banks and telecoms infrastructure.
- *Community Cloud.* Organizations sharing common business functions and/or objectives could collaborate and establish their own specific community Cloud infrastructure. Example of this include Associated Newspapers which is a group of newspapers and publishing media that establish a community Cloud infrastructure to serve their common needs.
- *Hybrid Cloud.* This deployment type is a mixture of private, community, and/or public Cloud. This is important to support higher resilience, availability, and reliability.

Public Cloud has many more customers than private and community Clouds. As a result, public Cloud hosts more services and has intensive interactions with customers. Managing the huge customer base of public Cloud necessitates the public Cloud only hosting services that could be fully managed automatically with minimal human intervention. Automation hides the complexity of the infrastructure and increases its resilience. At the current time fully automated management services are not yet available for most types of applications and virtual resources. Such a lack of automated management services forces public Cloud providers to mainly support basic services which can be automated. These basic services currently cover the needs of casual users, small businesses, and uncritical applications.

Community and private Cloud deployment types, however, establish strong relations with their customers. That is, customers typically have a relationship of mutual benefit or shared goals with the Cloud provider; customers may also be contractually bound to good behavior. These characteristics give rise to a substantial degree of trust in the Cloud; its architecture is also important, but perhaps less so. By contrast, users of public Clouds are much more reliant on infrastructure properties in order to establish trust.

The hybrid Cloud model is different from the above as it is a mixture of different Cloud deployment types. Carefully managing it could result in higher reliance, reliability or even a reduction in costs. For example, a hybrid Cloud could be composed of a public and private Cloud such that the private Cloud hosts the critical and dependent application and the public Cloud acts as a web front-end or stores protected backup. The hybrid Cloud could also result in higher risk if badly configured and managed. For example, if a hybrid Cloud is composed of a highly secure private Cloud and a public Cloud, an attacker could attack the weakest link (i.e., the public Cloud) and from there get into the private Cloud. Therefore, careful risk analysis and management would need to be conducted not only when outsourcing services into public and community Cloud types, but importantly when moving into a hybrid Cloud type.

1.6 Main Challenges of Clouds

The EU study of Cloud [3] states the following:

Cloud technologies and models have not yet reached their full potential and many of the capabilities associated with Clouds are not yet developed and researched to a degree that allows their exploitation to the full degree, respectively meeting all requirements under all potential circumstances of usage.

This strong statement implicitly indicates that Cloud is still at an early stage of development and there are lots of challenges that still need to be addressed in this domain. In this section we highlight the most commonly discussed challenges in Clouds. This book discuss the challenges and how they could be managed using today's technologies. The main challenges in Clouds are as follows:

- *Operational management.* The scale, heterogeneity, and number of services and users of Cloud computing are by far more complex than traditional enterprise infrastructure. This requires automating the process of managing the Cloud environment as the management of Cloud computing is beyond the capabilities of typical human administrations and current system management tools. Providing fully automated management services is one of the key challenges in Cloud, which is discussed in great detail throughout this book. The following are example of cases which currently require excessive human intervention:
  - *Automated and effective elasticity property.* This means that resources which are used by a service should reflect the real needs of the service. For example, running applications should immediately utilize allocated virtual computation, storage, and memory resources without the need to do further updates and/or restarts. This is not provided effectively at the current time, which results in an increase in operational management costs and,

in addition, affects the green agenda. Such a case would require an optimized scheduler which considers the green agenda, SLA and QoS. For example, it is more efficient to not power up resources and delay execution if (i) utilized resources will be available shortly and (ii) SLA/QoS are maintained.

- *Self-detection of failure and automated recovery.* Failure management within enterprise infrastructures is provided manually with support from the limited available management tools. Such a semi-automatic process reduces the resilience and availability of the infrastructural resources.

- *Data management.* The amount of stored data in the Cloud is huge and increasing massively. Controlling the distribution of data is a big challenge that requires full consideration of legislation, security, privacy, and performance factors. This problem is considered in the first part of this book. The following are examples of data management problems:
  - The huge volume of Cloud data affects data availability and transmission, as the greater the size of data the more complex it is to control its movement across the distributed elements of Clouds.
  - The lack of automated data management mechanisms has a direct effect on the provided QoS.
  - Data management is a major concern when scaling and shrinking resources, which is a result of Cloud elasticity. Cloud elasticity requires ensuring consistency and security of data when replicated and shrunk.
  - Classical DBMS may break in Cloud considering the latency of accessing disks and the cache coherency across a very large number of nodes.

- *Privacy, security, and trust.* Establishing trust in Clouds is the ultimate objective of most research in this domain. Other discussed challenges will eventually help in establishing trust in the Cloud. Privacy, security, and trust is a top challenge of Cloud that directly prevents its wider adoption, especially by critical infrastructure. Clouds suffer from major security concerns, for example: physical resources shared by many (possibly competing) customers – what is known as the *multi-tenant architecture*; vulnerability to the insiders threat of traditional enterprises; complex and heterogeneous architecture increasing security vulnerabilities. In addition, the Cloud elasticity results in security vulnerabilities when replicating, distributing, and shrinking data. This process must validate the non-existence of security holes in remote servers. Equally importantly, in current Cloud users do not have control over their resources, for example users cannot be assured about the way Cloud manages resources, about the integrity of their bills, and about Cloud's compliance with the greed SLA.

- *Forensic and provenance in Clouds.* This is one of the main issues in Clouds, and it helps in addressing many other challenges. A key fundamental requirement for establishing trust in Cloud is having a trustworthy provenance mechanism. Provenance helps in supporting proactive service management, assuring the integrity of bills, providing incident management, and lessening the impacts of insider threats, which increase Cloud trustworthiness. We devote Chapter 10 to this important topic.

- *Federation and interoperability.* The future vision of Cloud computing is to be the Internet-scale critical infrastructure. This strong vision requires trustworthy and resilient Cloud infrastructure that can survive even with failures of multiple Cloud providers. Addressing such a requirement requires establishing a Cloud-of-Clouds (what is also referred to as federated Clouds). The future vision of Cloud computing also enables customers to switch transparently between Cloud providers. Such visions (i.e., federated Clouds and flexibility in changing Cloud providers) are not available at the current time. One of the main reasons for this is the lack of standardization in this domain. The first part of this book presents the taxonomy of federated Clouds and briefly discusses this challenge.

- *Performance management.* This is a key subject for the success of Clouds, especially when considering the complexity, enormous customer base, and criticality of the Cloud. For example, high performance is a key for: managing the operation of the Cloud (e.g., scalability and resource scheduling), copying large amounts of data within the Cloud infrastructure and across federated Clouds, copying large amounts of data between Cloud customers and the Cloud infrastructure, and copying large amounts of data across distant locations within the Cloud infrastructure and across federated Clouds.
- *Legislation and policies.* Different countries have their own legislation in terms of where data could be hosted and which data is allowed. Cloud computing has many limitations for complying with different legislations. For example, current Cloud does not have the capabilitly to allow users to enforce the location of where their data could be stored and/or processed. In addition, current Cloud computing does not provide users with the capability to enforce their requirements (e.g., data privacy and security) and neither does it provide the assurance of their enforcement. This book does not cover the details of legal issues in Clouds; however, the frameworks which are discussed in the second part of the book look at how it addresses some of the legal requirements.
- *Economical aspects.* It is not always the case that switching to Cloud would provide the most economical approach. This is especially the case for well-established businesses that already have an enterprise infrastructure. Organizations would need to carefully balance and understand the risk and economical values when switching to Clouds. This book discusses the factors that would need to be considered when switching to Cloud, what services to outsource into Cloud, and the Cloud type that best suits an organization's needs.

Cloud computing helps in supporting green IT. For example, it offers possibilities to reduce carbon emission through more efficient resource usage; however, this needs to be counterweighed with the indirect carbon footprint arising from more experimental and thus more overall usage of resources, and the pressure on Cloud providers to update their infrastructure more regularly and faster than the average user.

1.7 Summary

Cloud computing is a recent term in IT, which started in 2006 with Amazon EC2. Cloud computing has emerged from commercial requirements and thus it draws an enormous amount of attention from both industry and academia, because of its promising future. Cloud comes with great advantages to help with economic growth, such as supporting the green agenda, reducing operational man-power, and providing effective utilization of resources. The lack of widely accepted academic studies that formally analyze the current Cloud infrastructure results in confusion over realizing its potential features, misunderstanding of some Cloud properties, and underestimating the challenges involved in achieving some of the potential features of Cloud. Discussing these was one of the main objectives of this chapter. The chapter also discussed Cloud services, deployment types, and main challenges. Subsequent chapters of the book build on the concepts presented in this chapter.

1.8 Exercises

Q1. What are the main features of Cloud which differentiate it from traditional data centers and enterprise infrastructures?

Q2. Cloud provides different services (i.e., IaaS, PaaS, and SaaS). Discuss the main differences between Cloud services.

Q3. Discuss the different Cloud deployment types.

Q4. What are the advantages and disadvantages of Clouds?

Q5. Organizations should understand the risks involved when outsourcing their data and services to public Clouds, and they should consider the available security and privacy options provided by Clouds. Can you identify some of the risks and how they could be managed?

Q6. The NIST definition of Cloud computing includes the statement 'minimal management effort or service provider interaction.' Discuss the importance of this statement in the Cloud definition.

References

[1] Michael Armbrust, Armando Fox, Rean Griffith, Anthony D. Joseph, Randy H. Katz, Andrew Konwinski *et al.* Above the Clouds: A Berkeley View of Cloud Computing. Technical Report No. UCB/EECS-2009-28, University of California, Berkeley, CA, February 2009.

[2] Peter Mell and Tim Grance. The NIST Definition of Cloud Computing, 2009.

[3] Keith Jeffery and Burkhard Neidecker-Lutz. The Future of Cloud Computing – Opportunities for European Cloud Computing and Beyond, 2010.

Part One

Cloud Management

2

Cloud Structure

This chapter presents the main components constituting the Cloud infrastructure. It also discusses relations and interactions between its components. Understanding the Cloud structure is vital to understanding Cloud properties, challenges, and the differences between its deployment types. This chapter also helps when discussing solutions for addressing Cloud problems.

2.1 Introduction

The Cloud infrastructure hosts various types of applications which could be simple, mid-range, or even highly complex. In addition, the Cloud infrastructure is accessed by a huge customer base. The huge number of applications hosted at the Cloud infrastructure, their variations, and the large customer base results in a highly complex and heterogeneous structure. Also, the differences in application requirements and the complexity of the infrastructure require Cloud components to be provided by different vendors. All these factors result in complexities in understanding the properties of the Cloud infrastructure and the relations between its entities. It gets even more complicated when considering the collaboration within a Cloud and across federated Clouds.

This chapter clarifies the structure of the Cloud and federated Clouds. Specifically, it focuses on the nature of Cloud resources, their grouping, types of data, and data flow across Cloud entities. Subsequent chapters of the first part of the book build on this chapter and clarify Cloud properties and management services.

This chapter is organized as follows. Section 2.2 briefly discusses the main components constituting Cloud infrastructure. Section 2.3 presents a 3-D view of Cloud computing and discusses the details of the grouping of the components within the Cloud infrastructure. Section 2.4 discusses all possible relations between Cloud components. Section 2.5 discusses the dynamic nature of Cloud. Section 2.6 discusses the types of data in the Cloud. Finally, we summarize the chapter in Section 2.7.

Cloud Management and Security, First Edition. Imad M. Abbadi.
© 2014 John Wiley & Sons, Ltd. Published 2014 by John Wiley & Sons, Ltd.
Companion Website: www.wiley.com/go/abbadi_cloud

2.2   Infrastructure Components

The Cloud infrastructure is composed of enormous components. High-level understanding of the functions of the Cloud infrastructure components, their properties and the way they interact is vital to understand Cloud computing. Cloud components have the following main categories: physical servers, storage components, network devices, and management platforms. This section briefly[1] discusses the functions of the first three categories while the last category is discussed in Chapter 3.

2.2.1   Storage Components

A storage component is a basic component[2] that stores Cloud data and/or provides file system services. Storage could be of two types: local storage and network storage. Local storage means that the storage component is connected directly to a server or multiple servers via a private network.[3] An example of this is the Storage Area Network (SAN) [2]. Accessing data at a local storage should be via a server component. Network storage, on the contrary, means servers are connected to a storage component over a public network.[4] An example of network storage is the Network Attached Storage (NAS) [3]. Network storage provides file storage as a service and therefore could be accessed directly by authenticated users and applications.

There are many important properties which are associated with the storage component, such as: size, speed, protection measures, and reliability. Enterprise architects are in charge of deciding on such properties when selecting and configuring a storage component. Such a decision would be based on the application properties that are planned to use the storage component.

2.2.2   Physical Servers

A physical server provides computational resources to Cloud users. It also provides possible means by which Cloud users could access network and storage resources. The server would typically run a hypervisor, which is a minimized operating system providing minimum components enabling the hypervisor to virtualize hardware resources to guest operating systems[4].

The hypervisor runs a Virtual Machine Manager (VMM). The VMM manages virtual machines (VMs) running at the physical server [4, 5] (e.g., starts, stops, and restarts a VM). A VM provides an abstraction of CPU, memory, network, and storage resources to Cloud users in such a way that a VM appears to a user as an independent physical machine. Each VM runs its own operating system (OS), which is referred to as guest OS. The guest OS runs its VM-specific applications. VMs running at the same physical platform would share the platform resources in a controlled manner but then should be independent and not aware of each other. For example, a VM can be shut down, restarted, cloned, and migrated without affecting other VMs running at the same physical platform.

---

[1] It is beyond the scope of this book to discuss the details of the Cloud components, apart from the management platform which is covered in detail. Our objective is to provide conceptual understanding of Cloud management rather than providing detailed understanding of every component in the Cloud.

[2] By basic component we mean an integrated component (e.g., EMC storage products [1]) and not a simple hard-disk or a physical block.

[3] A private network would typically connect storage to servers via a direct point-to-point cable or via dedicated switches.

[4] A public network, in our context, is not restricted to the Internet, as it could be an office LAN or WAN.

2.2.3   Network Components

The network of Clouds is the backbone which provides the communication medium between the resources constituting the Cloud infrastructure. There are many important properties associated with the network components, such as network speed, network nature, and restrictions affecting information flow as in the case of a firewall filtering traffic. Enterprise architects decide on the network properties.

The communication between Cloud resources is horizontally, vertically, or a combination of both. We define these as follows.

Horizontal communication. This is where Cloud resources communicate as peers. There are many examples of horizontal communication, such as replicating files between peers of virtual machines and synchronizing shared memory across parallel servers.

Vertical communication. This is where Cloud resources communicate with other Cloud resources following a process workflow in either up–down or down–up directions. This would typically work as follows. First, an upper layer's resource runs a process which generates sub-processes that must be run at lower layers. The lower layer would then process the sub-processes and send the outcome to the upper layer. These steps represent an up–down communication channel. Each layer in turn sends their response back in the opposite direction, which represents the down–up communication channel.

2.3   Cloud Layers

The Cloud infrastructure is analogous to a 3-D cylinder, which can be sliced horizontally and/or vertically (see Figure 2.1). We refer to each slice using the keyword 'layer.'

A layer represents Cloud resources that share common characteristics.

The layering concept helps in understanding the relations and interactions amongst Cloud resources. We use the hosting relation between resources as the key characteristic for horizontal slicing of Cloud (i.e., physical, virtual, or application). We use the function of the resource (i.e., server, network, or storage) as the key characteristic for vertical slicing of Cloud. Figure 2.1 illustrates the 3-D view of the Cloud. The side view of the Cloud results in horizontal slices and the top view results in vertical slices. The following subsections discuss these views.

2.3.1   Vertical Slices

As illustrated in Figure 2.1(b), the top view of the Cloud results in three layers (that is, by considering the function of resources): a storage layer, a server layer, and a network layer. As the names indicate, the storage layer consists of storage components, the server layer consists of physical servers, and the network layer consists of the network components. Unlike the side view of Clouds, the top view is not concerned about software stacks inside these layers.

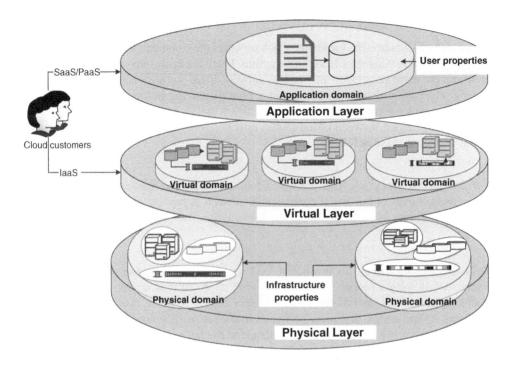

(a) Horizontal slice/side view

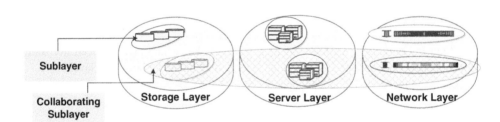

(b) Vertical slice/top view

Figure 2.1   Cloud taxonomy: 3-D view

The components of each layer are organized into three groups: network, storage, and server. We refer to the groups within a layer as sublayers; that is, we have network sublayers, storage sublayers, and server sublayers. A sublayer, for example, could be a cluster of physical servers, a replicated integrated-storage system, or a set of high-availability switches. The properties of the members of each sublayer are carefully selected such that a sublayer can satisfy its planned properties. A server sublayer is then connected to a storage and a network sublayers to form a unified group, which is called a collaborating sublayer. The associations between the three sublayers are not random. In fact, it must not be random as it is based on a careful enterprise architecture design. The architecture is meant to provide certain attributes enabling it to host certain categories of applications. The collaborating sublayer has properties which are inherited directly from its member sublayers. Such properties enable it to serve the needs of a common category of user application requirements (e.g., a database management system that supports transaction survivability). For example, when a user interacts with a Cloud provider to host a VM, the user would instruct the provider to create a VM with a set of requirements. The Cloud provider would then host the VM using a collaborating sublayer, which has properties that can meet the user requirements.

2.3.2   Horizontal Slices

As illustrated in Figure 2.1(a), a side view of Cloud computing results in three layers (that is, by considering the hosting relation between resources): a physical layer, a virtual layer, and an application layer. The *physical layer* represents physical resources and their management software components.

The management software component could, for example, be an operating system, a hypervisor, or a firmware. The physical layer is associated with policies and properties. The policies and properties govern the interactions of physical resources and are used when managing and enforcing the requirements of users at the physical layer. A physical resource could be of server type, storage, or network device. The physical layer should be completely transparent to Cloud customers.

The virtual layer consists of virtual resources which are hosted at physical resources. The virtual resource could be a virtual machine, a virtual disk, or a virtual network, which collaboratively serve software platforms and applications at the application layer. Cloud customers using IaaS have to fully maintain their VMs including guest operating systems without worrying about the resources of the physical layer. The virtual layer should be transparent to Cloud customers using SaaS service, and possibly PaaS service.[5]

The *application layer* represents resources of software applications and software platforms (herein we refer to application resources for simplicity to mean either a software application resource or a software platform resource). An application resource could be hosted either at virtual resources or directly at physical resources. The earlier is the most commonly known case, that is hosting an application resource at the virtual layer. The latter case is also valid in a Clouds context and is the common case for both critical infrastructure and highly resilient application requirement. That is, Clouds could host applications directly at a physical layer without necessarily having a virtual layer.

The resources of the horizontal slices are categorized and grouped based on their types, properties, functions, and deployment across the Cloud infrastructure. Understanding such grouping and their associated properties helps us understand where in Clouds user requirements need to be enforced. Entities within a layer are organized into *domains*. That is we have physical domains, virtual domains, and application domains. A domain represents related resources which enforce the domain policy. The physical domains are related to a specific Cloud infrastructure and, therefore, are associated with the infrastructure properties and policies. The virtual and application domains are a Cloud user specific and therefore are associated with the Cloud user properties.

---

[5] The level of transparency would depend on customer settings as a customer could have a combination of IaaS and PaaS.

Domains that need to interact amongst themselves within a layer join a *collaborating domain*, that is we have physical collaborating domains, virtual collaborating domains, and application collaborating domains. A collaborating domain controls the interaction amongst its members using a defined policy. Domains and collaborating domains help in managing the distribution and coordination of the resources of the Cloud infrastructure in normal operations as well as during incidents such as hardware failures or a whole data-center failure. The exact purpose of such domains is layer specific, as discussed in Section 2.4.

Federated Cloud is a new concept in IT (also called Cloud-of-Clouds). It is proposed to enable different Cloud providers to collaborate in emergencies. Examples of an emergency include acts of God (e.g., an earthquake), site failure, or even when a Cloud provider exceeds its capacity limits. Federated Clouds are proposed to enable Cloud to be the Internet-scale critical infrastructure resembling the Internet backbone. Managing federated Clouds to a certain extent is similar to managing collaborating organizations and somehow similar to managing a geographically distributed Cloud infrastructure. However, in the federated Cloud case more stringent restrictions need to be applied.

Collaborations within the federated Cloud can be abstracted in the form of *emergency domains* (see Figure 2.2). An emergency domain consists of a set of physical collaborating domains where each collaborating domain is specific to a Cloud provider. In this, the participating Cloud providers of each emergency domain should be members of a federated Cloud. The selection of the members of an emergency domain should be planned in such a way that the members should be capable of serving as backup for each other without compromising user requirements.

Emergency domains are associated directly with the physical layer. Virtual and application layers do not have the emergency domain concept directly; however, their resources would benefit indirectly from federated Clouds when they are hosted at a physical collaborating domain which is itself a member of an emergency domain. That is, application and virtual resources would inherit the properties of their allocated physical collaborating domains, as clarified in Section 2.4. If the physical collaborating domain is a member of an emergency domain, then all the resources which are hosted at the physical collaborating domain would inherit the benefits of the emergency domain.

2.3.3   Horizontal vs. Vertical Slices

The top and side views of the Cloud infrastructure are important to understand the properties of the infrastructure, the relation between its resources, and the way they are managed and organized. The views are not meant to replace each other. It is rather the opposite, as these views present different angles of the Cloud. The way to view the Cloud depends on the nature of the situation being considered. For example, the components of the physical infrastructure and their properties are better realized by the top view; however, the hosting relation between Cloud resources and the communication between different types of resources are realized using the side view. In some cases both the side and the top view would be needed to understand the scope of a situation, as in the case of managing disaster recover scenarios. Subsequent chapters of this book provide detailed examples of the different views of Cloud.

The following is a high-level comparison between both views. The side view of the Cloud does not have the sublayer concept. The collaborating sublayer concept in the top view is

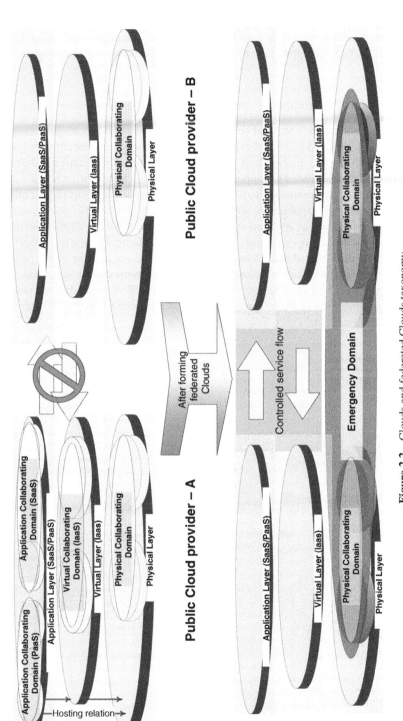

Figure 2.2 Clouds and federated Clouds taxonomy

similar to the concept of a physical collaborating domain in the side view. The top view of the Cloud does not have the concept of a collaborating physical domain. Neither does it discuss the grouping of virtual and application resources into domains and collaborating domains.

2.3.4   Illustrative Example

Figure 2.3 illustrates the side view of the Cloud using a simplified multi-tier example. The physical layer is composed of two physical domains: the first is physically configured and optimized by a Cloud provider to host DBMS applications which can provide no single point of failure (e.g., configured to support a real application cluster [6]), while the second is physically configured and optimized to host lightweight-type applications (e.g., a middle-tier application running a web-application server).

The application layer has an application domain which consists of two application resources: a middle-tier application component and a backend DBMS. The application resources represent a multi-tier application hosted at a Cloud infrastructure. The application owner provides his requirements for each application resource (e.g., minimum/maximum scalability value of each resource, physical location restrictions, and backend DBMS providing no single point of failure). An application resource could be hosted at the virtual or physical layer based on

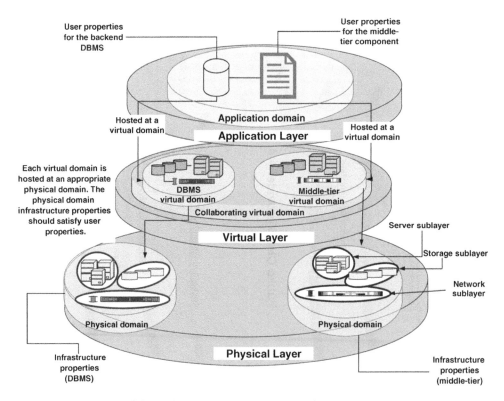

Figure 2.3   Multi-tier example using the provided Cloud structure

the application requirements. The application domain manages the hosting of its resources at appropriate virtual or physical resources. For example, a DBMS resource would typically be hosted at the physical layer without involving virtual resources. In this case, the application domain would manage the hosting of the application backend resource at an appropriate physical domain. In contrast, the middle-tier application resource would typically require being hosted at the virtual layer. In this case the application domain would interact with the virtual layer to create a virtual domain. The created virtual domain would in turn manage the hosting of the virtual resources at an appropriate physical domain following the application owner-requested requirements.

The concept of the collaborating virtual domain represents a group of virtual domains in which each virtual domain depends on the others. Resources of dependent virtual domains could be related to the same application, or they could be related to different applications. For example, an organization might have multiple applications which exchange a high volume of transactions. That is, such types of applications, for performance reasons, need to run within physical proximity. This is achieved by joining the virtual domains where the applications run to a collaborating virtual domain and then associating the collaborating virtual domain with properties and policies. Such policies and properties would mainly be based on user requirements and they would help the Cloud provider to automatically manage the group of domains based on these requirements. This will be clarified further in subsequent chapters.

The concept of the collaborating physical domain can be realized by having redundant physical domains which have similar infrastructural properties; for example, a collaborating domain in our scenario requires creating a new physical domain with the same properties as the DBMS physical domain, and then joining both domains within a collaborating physical domain. In this case, the members of the collaborating physical domain (based on a defined policy) would automatically act as backup for each other during emergencies.

2.4   Cloud Relations

This section clarifies the interactions and dependencies of resources and groupings within a layer and across dependent layers. The dependencies and interactions are very well organized and managed following predefined policies. Such policies are based on both user requirements and infrastructure properties and policies. The relations between Cloud resources are of two main types: intra-layer and across-layer relations, which are discussed in this section.

2.4.1   Intra-layer Relations

Intra-layer relations exist within the same layer of either a single Cloud or across federated Clouds. We identify the following relations that fall under this category, as illustrated in Figure 2.4.

- *Resource–Resource (R-R-IL)*. Resources within a single layer communicate amongst each other to provide services to other layers. This relation can exist within a single domain or across several domains. Resources should always adhere to their domain policy.
- *Resource–Domain (R-D-IL)*. Relation covers the management of a resource by a domain which could be of two types: resources communicate with their own domain or resources

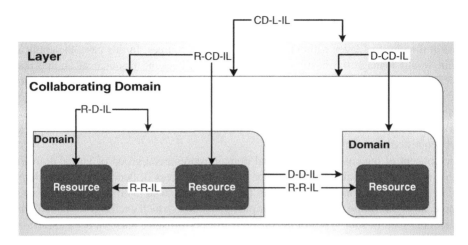

Figure 2.4   Intra-layer relations

communicate with other domains. We are mainly interested in the first case since, in the
Cloud context, a resource is managed based on its specific-domain policies. This relation
is bi-directional in that the behavior of a resource may be influenced by the behavior of
the domain it is a member of. In the other direction, the behavior of the domain is directly
influenced by the behavior of its resources.

- *Resource–Collaborating Domain (R-CD-IL)*. Relation is used to manage resources in
  exceptional circumstances. A resource communicates and is managed by its own domain,
  and a domain provides Cloud properties (such as high resilience) by communicating
  with other domain members of the same collaborating domain. As a result, the relation
  R-CD-IL is transitive and covered under the following intra-layer relations: R-D-IL and
  D-CD-IL.
- *Domain–Collaborating Domain (D-CD-IL)*. Relation is bi-directional in such a way that
  the behavior of a domain may be influenced by the policies set in its specific collaborating
  domain. In the other direction, the behavior of a collaborating domain is directly influenced
  by the status of the domains which are members of the collaborating domain. In addition, if
  the collaborating domain is a member of an emergency domain, then its behavior is directly
  influenced by the policies of the emergency domain. This relation shows a view of the
  interdependencies between the resources of a specific Cloud customer.
- *Domain–Domain (D-D-IL)*. Relation represents the interdependencies between a group of
  domains. These could either exist within a specific collaborating domain or across various
  collaborating domains. Within a 'single Cloud' we restrict domain relations to be within
  a collaborating domain and insist that if a domain depends on another domain, then both
  domains must be members of the same collaborating domain. On the contrary, the 'federated
  Clouds' case requires physical domain dependencies to be, in addition, across various
  physical collaborating domains. In this case, the collaborating domains involved must be
  members of the same emergency domain and must belong to different Cloud providers.
  Policies associated with emergency domains govern the interactions of member domains
  during emergencies.

- *Collaborating Domain–Emergency Domain (CD-ED-IL)*. Relation represents dependencies between a set of collaborating domains with their emergency domain. This relation is only valid for physical collaborating domains. Specifically, it governs the behavior of Clouds during emergencies which require the involvement of other public Clouds that form a federated Cloud. Resources should not be aware of the emergency domain. During emergencies, the collaborating domains would follow the rules of their emergency domain policy.
- *Collaborating Domain–Layer (CD-L-IL)*. Relation represents a membership relation. In contrast to the other types of membership relations, this relation is one-directional in that a collaborating domain does not influence the behavior of the layer, instead it only represents a 'slice' of the resources provided in a layer, that is a collaborating domain is conceptual as far as layers are concerned.

2.4.2   Across-layer Relations

Across-layer relations exist between entities in different layers, as illustrated in Figure 2.5.

- *Resource–Resource (R-R-AL)*. A resource in one layer communicates with resources that are members of other layers for hosting and message communication purposes. Examples of this relation include: the hosting relation between application, virtual, and physical resources, and the communication between dependent resources across layers as in the case of the communication of an application resource with dependent virtual resources.

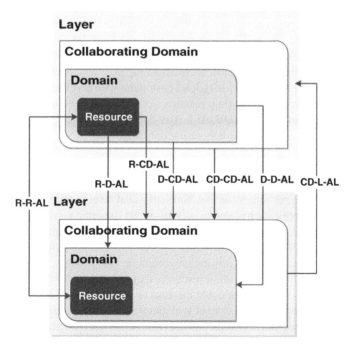

Figure 2.5   Across-layer relations

- *Resource–Domain (R-D-AL)*. Relation resembles the management of a resource in one layer by a hosting domain at another layer. For example, a virtual domain manages the hosting of an application resource; analogously, a physical domain manages the hosting of a virtual resource or an application resource (based on the hosting relations discussed earlier).
- *Resource–Collaborating Domain (R-CD-AL)*. Relation follows the same principle as R-CD-IL.
- *Domain–Domain (D-D-AL)*. Relation is used mainly to manage domain interdependency across layers. It is related to the management of the resources hosting environment, which covers the following cases: an application domain might be hosted by at least one virtual domain; a virtual domain might host an application domain or a specific resource of an application domain; a virtual domain should be hosted by one or more physical domains; a physical domain might host multiple virtual domains; a physical domain might host the resources or part of the resources of an application domain; an application domain might be hosted by one or more physical domains.
- *Domain–Collaborating Domain (D-CD-AL)*. Relation is layer-specific as follows: the D-CD-AL relation between an application or a virtual layer and a physical layer is used mainly to manage backup domains at the physical layer (e.g., if a physical domain fails then its hosted resources could be relocated to another backup domain member of the same collaborating domain); the D-CD-AL relation between an application layer and a virtual or a physical layer is used mainly to manage the interdependencies amongst application resources, for example performance requires application resources which have transaction rates to be within physical proximity to reduce network latency.
- *Domain–Emergency Domain (D-ED-AL)*. Relation follows the same principle as D-CD-AL.
- *Collaborating Domain–Collaborating Domain (CD-CD-AL)*. Relation represents the overall Cloud customer-dependent resources within the Cloud environment. For example, dependent application resources will be within an application collaborating domain which is hosted by a set of virtual or physical collaborating domains. Similarly, a virtual collaborating domain is hosted by a set of physical collaborating domains. As a result, understanding CD-CD-AL across all layers provides Cloud customers with a vertical and horizontal slice of the Cloud infrastructure where their resources could possibly be managed and hosted.
- *Collaborating Domain–Layer (CD-L-AL)*. A collaborating domain is not related to other layers, as each layer has its own collaborating domains.

2.5 Cloud Dynamics

The building up of Cloud resources starting from a physical resource hosting a virtual resource which, in turn, runs an application resource is dynamic. By dynamic we mean the following: (a) a specific virtual resource has a $1:N$ relation with a physical resource, that is a specific virtual resource can run on different physical resources following a predefined policy restricting and controlling the virtual resource hosting environment; (b) similarly, a specific application resource has a $1:N$ relation with a virtual resource, that is a specific application resource can run on multiple virtual resources that could increase or decrease based on a predefined policy controlling such behavior (as in the case of horizontal scalability, covered in Chapter 5); and (c) from points a and b we can conclude that the relation between a specific application and a physical resource is $1:N$, that is a specific application can eventually be hosted under different physical servers.

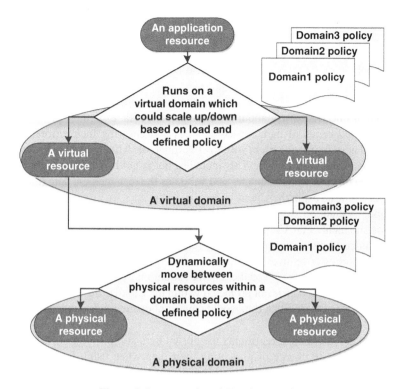

Figure 2.6 Dynamics of Cloud computing

Points a, b, and c show that the building block of a Cloud environment is not static but dynamic. This is not to say that a Cloud resource would run anywhere on the Cloud's physical infrastructure, it is rather the opposite, as Cloud resources are well controlled and managed following policies controlling the limits of movement [7]. That is, an application resource can move within a specific virtual domain boundary, and similarly a virtual resource can move within a physical domain boundary, as discussed in previous subsections. We illustrate this dynamic relation in Figure 2.6.

The dynamic nature of Clouds has tremendous advantages for enhancing their desirable properties, as in the case of resilience, availability, and scalability (discussed in Chapter 4). However, it is also associated with security challenges that are unique to this environment, as covered in Part Two of the book.

2.6 Data Types

There are two types of information flow in Clouds: one holds management data and the other holds application data. Management data is related to the information which is needed to manage physical, virtual, and application resources and their interactions. The management data is not about Cloud's management services, which we discuss in Chapter 5; however, it covers all the data which the management services would require in order to operate the Cloud infrastructure and it also covers the outcome of such services. This, for example, includes the data for managing the hosting of virtual resources at physical resources, and application resources at virtual or physical resources. In addition, the management data covers the data for managing resource groupings and their interactions within physical, virtual, and application domains and collaborating domains. Moreover, it covers issues of federated Clouds such as managing emergency domains and their policies.

Cloud providers are in charge of managing the 'management data.' Cloud users, on the contrary, shouldn't need to worry about the management data, except for providing that representing the requirements of their hosted services and ensuring their enforcement. Cloud providers would need to provide users with trustworthy tools which help them to transparently assess the trustworthiness of Clouds. This includes, but is not limited to, assessing the trustworthiness of the process which manages and validates the management data (the second part of this book discusses trust in the Cloud).

The continual validation, verification, and protection of management data is crucial for a successful management of the Cloud infrastructure, for example, help in proving the assurance that the Cloud infrastructure is operating as expected, providing proactive service delivery, and assuring and supporting error investigation processes. Unlike application data, the protection of management data has not yet received the required attention from both industry and academia. For example, management data is not protected within the Cloud infrastructure and, as a result, it is subject to insider attack. Such data could cause unbounded consequences and should be protected by Cloud providers.

Application data, in contrast, represents the hosted data at the Cloud infrastructure. Users of the hosted data are in charge of protecting their data when using IaaS and PaaS Cloud services. For SaaS, the Cloud provider should ensure the protection of user data. Users should be provided with trustworthy tools to help them in this process, such as key management as a service and provenance as a service. For federated Clouds, case users should not be required to use special methods when accessing and protecting their data. That is, application data in federated Clouds should be managed, from a customer's perspective, in the same way as if it was hosted only within a single Cloud. For example, when a customer is encrypting, storing, and accessing their data they should not need to worry about data replication across federated Clouds, the variations in the hardware used or changes in the underlying hypervisor, or the identity of the Cloud provider which served their applications data.

Cloud providers should not be able to access their customers' data; however, in order to present a transparent infrastructure management, Cloud providers should still have a fine line of control over part of the management of the applications data. Such partial management should not remove control of the application data from the hands of customers to the hands of the Cloud provider. The partial management is mainly for managing the hosting environment of the software applications. Managing the hosting environment is related to managing the availability, scalability, and resilience of application data by considering the user requirements provided. For example, during emergencies, customer data should be inaccessible for a longer period than the agreed time in the SLA. Cloud management services use the management data to provide such transparent management of the application data.

2.7 Summary

This chapter has presented the Cloud ontology. It started by discussing the main components of a Cloud infrastructure. It then presented the 3-D view of Clouds, which is composed of horizontal and vertical slices. The 3-D view helps in clarifying the structure of the Cloud and the grouping of its entities. The chapter discussed the relations amongst Cloud resources and grouping of resources. Finally, it discussed the data types of Clouds.

The presented ontology of Clouds is the foundation which other chapters build on. For example, it helps in understanding the infrastructure properties of Cloud, how to enforce user requirements without losing control, and importantly how to establish trust in Clouds. In addition, the Cloud ontology helps in realizing the complexities of providing self-managed services. These subjects are covered throughout the book.

2.8   Exercises

Q1. What are the advantages of the Cloud taxonomy?

Q2. What are the differences between the top and side views of Clouds?

Q3. Identify and discuss the main categories of Cloud components.

Q4. What is the dynamic nature of Clouds?

Q5. What are the pros and cons of the dynamic nature of Clouds?

Q6. Discuss the differences between Cloud management and application data.

References

[1] EMC. 2012. http://www.emc.com/products/category/storage.htm.
[2] Wikipedia. Storage Area Network (SAN), 2010. http://en.wikipedia.org/wiki/Storage_area_network.
[3] Wikipedia. Network-Attached Storage (NAS), 2010. http://en.wikipedia.org/wiki/Network-attached_storage.
[4] Derek Gordon Murray, Grzegorz Milos, and Steven Hand. Improving Xen security through disaggregation. In *Proceedings of the Fourth ACM SIGPLAN/SIGOPS International Conference on Virtual Execution Environments, VEE '08*, pp. 151–160. ACM: New York, 2008.
[5] Jonathan M. McCune, Yanlin Li, Ning Qu, Zongwei Zhou, Anupam Datta, Virgil D. Gligor, and Adrian Perrig. Trustvisor: Efficient TCB reduction and attestation. In *IEEE Symposium on Security and Privacy*, pp. 143–158, 2010.
[6] Oracle. Oracle Real Application Clusters (RAC), 2011. http://www.oracle.com/technetwork/database/clustering/overview/index.html.
[7] Imad M. Abbadi. Toward trustworthy clouds' internet scale critical infrastructure. In *ISPEC '11. Proceedings of the 7th Information Security Practice and Experience Conference*, vol. 6672 of *LNCS*, pp. 73–84. Springer-Verlag: Berlin, 2011.

3

Fundamentals of Cloud Management

This chapter presents fundamental concepts about Cloud management platforms. Understanding the Cloud management process helps in realizing Cloud attributes, which are key for the development and enhancement of Cloud computing. The chapter starts by identifying and discussing the main services for managing Cloud resources. It then presents a unified view of the most well-known Cloud management platforms. Subsequently, the chapter discusses the required input data that such platforms would need to consider in order to make the right decisions. Following that, the chapter presents an end-to-end workflow for managing user requirements in the Cloud. We then identify the weaknesses in the workflow. Finally, we identify and discuss the challenges of managing the Cloud infrastructure, and present the requirements which could address the challenges.

3.1 Introduction

Based on the NIST definition of Cloud computing (see Chapter 1), one of the key features of Cloud is the provision of minimal management effort or service provider intervention. This is a key attribute of Clouds, and without it the Cloud cannot practically satisfy many of its potential features for different reasons, as we discuss throughout this chapter. Meeting this attribute is a major challenge in the Cloud environment, and would require enormous collaborative efforts between academia and industry. The provision of such a feature is not only important to reduce management costs and overhead, but is also one of the key requirements for establishing trust in the Cloud. This chapter clarifies the management aspect of Cloud and the importance of supporting Cloud with self-managed services. Subsequent chapters build on this chapter and discuss further issues in Cloud management and security.

Cloud Management and Security, First Edition. Imad M. Abbadi.
© 2014 John Wiley & Sons, Ltd. Published 2014 by John Wiley & Sons, Ltd.
Companion Website: www.wiley.com/go/abbadi_cloud

The chapter is organized as follows. Section 3.2 identifies the required management software services of the Cloud infrastructure. Section 3.3 analyses Cloud's management platforms of virtual resources. Section 3.4 discusses the main input data that would affect decisions made by the identified management software services. Section 3.5 provides a detailed analysis of the overall process for managing user requirements inside the Cloud infrastructure. Finally, we summarize in Section 3.6.

3.2   Clouds Management Services

Supporting Cloud with self-managed services is the foundation for providing trustworthy Cloud infrastructure that is capable of becoming the Internet-scale critical infrastructure. This section identifies these services and starts by providing a real-life application deployment in Cloud. Subsequently, we use the scenario to identify the required management services and motivate the importance of such services.

3.2.1   Application Deployment Scenario

The scenario discussed in this section is based on a real-life deployment of a system in a production environment supporting an editorial workflow. This system is composed of an editorial application and a weather forecasting application. For simplicity, we assume both editorial and weather applications have similar architectural requirements. The system is architected as a multi-tier application which is deployed across a community Cloud infrastructure. The Cloud infrastructure is split across two locations: primary and secondary, as illustrated in Figure 3.1. Having an infrastructure that is physically distributed at different geographical locations helps Cloud in providing disaster recovery (DR), supporting business continuity, and increasing Cloud availability, reliability, resilience, and scalability. We now provide a simplified architecture of the deployed application scenario, as illustrated in Figure 3.1.

The system has two application domains: an editorial application domain and a weather application domain. Each application is composed of layers: a middle-tier layer and a backend layer. The middle-tier layer for each application domain runs the business process and is hosted at a specific virtual domain, while the backend layer of each application domain manages the application database and is hosted directly by physical domains. This fits with the discussed Cloud taxonomy in Chapter 2.

The system requires two virtual domains: one for hosting the weather application middle-tier layer and another for hosting the editorial application middle-tier layer. The number of VMs for each virtual domain and their specifications would depend on application requirements as agreed in the SLA.[1] In general, each domain should have at least two VMs to support higher resilience with no single point of failure. Each backend application domain resembles a database instance. The database instance would be hosted at a database management system (DBMS). This would typically be deployed directly at an appropriate physical domain.[2]

---

[1] We do not discuss application requirements in the example for simplicity.
[2] Chapter 2 discusses the pros and cons of application hosting at virtual and physical domains.

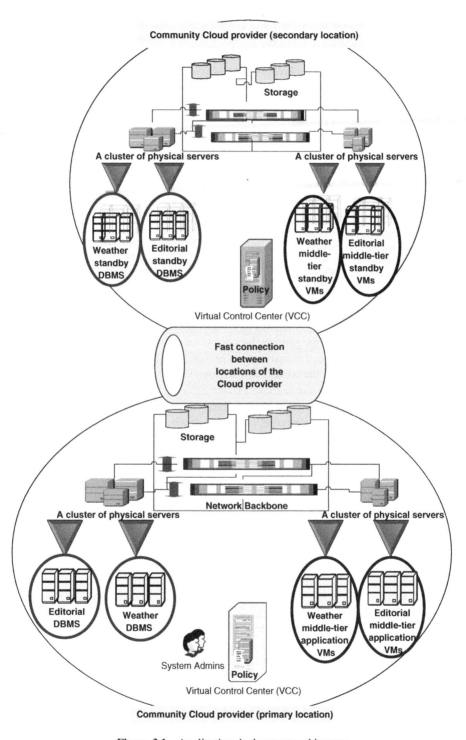

Figure 3.1   Application deployment architecture

The system requires at least two types of physical domains: one to host a middle-tier type of application and another to host a DBMS. Each domain would be composed of a storage sublayer, a network sublayer, and a server sublayer. Each sublayer has special properties, and the collaborations amongst the sublayers would provide the physical domain with properties that enable the domain to host the indicated type of application. At this level the focus is mainly on non-functional requirements including resilience, availability, reliability, adaptability, scalability, security, and privacy.

The higher the resilience requirement the more complex the architecture would be. In our example, as illustrated in Figure 3.1, the system is hosted at a highly resilient component at the physical infrastructure. The architecture of the technology requires each physical domain to be a member of a collaborating physical domain such that the members of the collaborating physical domain are hosted at different geographical locations. This is implemented by hosting the first physical domain at the primary site and the second physical domain at the secondary site, as follows:

- A middle-tier (primary) physical domain (a) has properties enabling it to host middle-tier application and is physically located at the primary location.
- A middle-tier (secondary) physical domain (b) has properties enabling it to host middle-tier application and is physically located at the secondary location.
- A DBMS (primary) physical domain (c) has properties enabling it to host DBMS instances and is physically located at the primary location.
- A DBMS (secondary) physical domain (d) has properties enabling it to host DBMS instances and is physically located at the secondary location.

The scenario requires four physical domains, which are related as follows: the physical domains a and b are members of one collaborating physical domain, and the physical domains c and d are members of another collaborating domain. The physical domains located at the secondary location act as backup for the physical domains located at the primary location. One of the roles of the enterprise architect team is to define the attributes of the physical domains to be capable of hosting software applications considering user requirements.

Understanding the nature of the hosted application enables enterprise architects to use enhanced features in terms the right hardware configuration that best suits the generic nature of the software application. For example, if application activities are more write than read then, for performance reasons, using RAID 1+0 for storage configuration is much better than RAID 5. Also, user requirements that indicate no single point of failure would imply an integrated storage component which should be fully redundant from inside and outside (e.g., dual communication channels and multiple processor cards). It also implies replicating data across different geographical locations, that is across the community Cloud primary site and secondary site. Replicating data can be done at different levels: storage managed replication and/or DBMS managed replication.

In our scenario, enterprise architects would, in addition, do the following:

- Associate the two application domains with dependency property, which necessitates the active copies of the application resources running within physical proximity. That is, emergencies that affect a single application resource would require all dependent application resources to fail-over to the secondary location. This, for example, would avoid the case where a DBMS application resource is hosted at a different location than its corresponding middle-tier application domain.

- Host editorial and weather DBMS domains at a physical domain that has a server sublayer with properties enabling it to host DBMS with no single point of failure.
- Host editorial and weather middle-tier virtual domains at a physical domain that has a server sublayer with properties enabling it to host middle-tier-type applications.

The storage and server sublayers must be connected using redundant network sublayers to support high-resilience architecture. For example, a DBMS server sublayer should be connected using multiple channels to the related storage sublayer.

Enterprise architecture is a complex subject which is beyond the scope of this book to discuss in further detail. Our main goal is to briefly present an example of an architected application to help in identifying Cloud attributes and management services.

3.2.2   Identifying Cloud Management Services

This section aims to identify the main services that the Cloud infrastructure supports to meet part of the applications' non-functional requirements. It also clarifies the importance of automating the management of these services. Automating the process of managing Clouds is one of the key requirements for moving critical applications to public Clouds. Critical applications are hosted mainly by private and community Cloud providers. Such applications, at the time of writing, cannot be hosted by public Cloud for several reasons. These are mainly related to the lack of automation and high reliance of humans. Enhancing Cloud with automated managed services is the foundation for establishing trustworthy Clouds.

Most of the Cloud management activities at this time still require excessive human interaction. Public Cloud providers have huge and well-structured IT departments. The IT departments have groups of employees with different roles. Each group is in charge of managing certain aspects of the Cloud, for example an architecture group, a deployment group, a system administration group, and a security management group. Such groups manage the Cloud infrastructure. In this section we identify the services that are provided by Cloud groups to help in drawing the required Cloud automated management services.

Automating the process of managing Cloud resources requires a set of self-managed services. We need to understand the process workflow of Clouds in order to identify these services. A simplified system architect is outlined in the previous subsection. The process workflow starts with the user, who first provides a set of requirements to a Cloud provider. The requirements are then verified and validated. The enterprise architect role provides a resilient architecture that satisfies the provided user requirements. This role is reflected in the following self-managed service: *system architect as a virtual service*. Next, the system administrator role deploys the architected system. This role is reflected in the following self-managed services: *resilience as an application service* and *resilience as a virtual service*.

Applications running at the Cloud infrastructure could fail for any reason. Also, users could change their requirements, and the Cloud provider could update the properties of the physical resources. Cloud providers, as a result, should automatically adapt to such events (i.e., unexpected failures, changes in user properties, and changes in infrastructure properties and policies). The roles that perform the adaptability activities in current Cloud provider models are referred to as incident and change management.[3] These roles could be application resource related and/or virtual resource related. Thus, there is a need to have the following self-managed services that can perform these roles: *adaptability as a virtual service* and *adaptability as an application service*. The adaptability service is critical for the potential Cloud infrastructure. For example, when users change their requirements, the virtual layer resources should automatically adapt to such changes; also, when the infrastructure physical resources get changed, the virtual layer resources should also automatically adapt to such changes without compromising user requirements. All these changes should not compromise user requirements, and security and privacy properties.

Elasticity (also referred to as scalability) is one of the most desirable properties of Clouds. In peak periods the virtual layer resources should automatically scale up, and at off-peaks the resources should automatically scale down. Such scaling is based on demand and is subject to customer pre-agreed SLA. Elasticity should not compromise user requirements, and security and privacy properties. At the time of writing, the major work behind Cloud elasticity is provided by the role of system administrators. Therefore, a self-managed service is needed to support elasticity, which is referred to as: *scalability as a virtual service* and *scalability as an application service*.

Other important services which are associated with elasticity are *availability as a virtual service*, *availability as an application service*, *reliability as a virtual service*, and *reliability as an application service*. The availability services ensure that all access paths to a resource are always available to requesters. For example, when a resource is replicated, the requester should automatically get notified about the availability of additional access paths. Similarly, when a resource gets shrunk, the requester should automatically get notified about the removal of an access path. The reliability services, however, ensure an end-to-end service integrity. Self-managed integrity assurance is different from the definition of integrity in information security. The latter provides the assurance that data has not been manipulated but does not prevent it happening. However, the former prevents data manipulation by malfunction process, malicious insider, attacher, etc.

Self-managed services should consider security and privacy by design. That is, the role of security professionals should not resemble an independent service; it is rather the opposite, as security and privacy should be considered fundamental parts within each service.

At the time of writing, the importance of self-managed services is recognized; however, little work has been done in this direction. This is because such services are complex and rely on huge collaborative efforts between industry, and industry and academia. For example, we still do not have global standards amongst different suppliers of many technologies. This is one of the keys for self-managed services. Enterprises are still highly reliant on human beings, who which require a longer time to architect and deploy solutions, and a longer time to discover and resolve problems. Also, reliance on human beings is error prone, subject to insider threats by Cloud employees, and does not provide a reliable way of measuring

---

[3] Depending on the complexity of the organization, the incident and change management could be one role, two separate roles, or part of a general role.

the level of trust in Cloud's operations. This raises the need for self-managed services that can automatically and with minimal human intervention manage the Cloud infrastructure. Automated self-managed services provide Cloud computing with exceptional capabilities and new features, such as: scale per use, hide the complexity of the infrastructure, automated higher reliability, availability, scalability, dependability, and resilience that considers users' security and privacy requirements by design. Automated self-managed services should help in providing trustworthy resilient Cloud computing, and should result in cost reduction. We discuss these services in further detail in Chapter 5.

3.3   Virtual Control Center

The Cloud infrastructure is composed of enormous and complex resources. Such resources cannot be efficiently and reliably managed only by relying on humans and basic management tools. Cloud management requires advanced and complex automated management tools. Currently, there are different tools which help administrators in managing the Cloud infrastructure. These include tools for virtual resource management, physical resource management, network management, cluster management, and database management. This book is mainly concerned with the virtual resource management tools, which could either be open-source or proprietary tools. Examples of propriety management tools include vCenter from VMWare [1], and System Center from Microsoft [2]. Example of open-source tools include OpenStack [3] and OpenNebula [4]. For convenience, this book refers to such tools using a common name: *Virtual Control Center (VCC)*.

A VCC is a Cloud device that manages virtual resources and their interactions with physical resources using a set of intelligent self-managed software agents.

The management of virtual resources requires the VCC to establish communication channels with physical servers. The VCC currently establishes the communication channels via a VMM. VMMs are installed at physical servers to manage their allocated VMs. Such management would follow the instructions received from the VCC, as illustrated in Figure 3.2. Current implementations of VCCs require each VMM to regularly communicate its VMs status (e.g., failure, uptime, and shutdown) to the VCC. The VCC and VMMs regularly exchange 'heartbeat' signals ensuring they are up and running. These enable the VCC to be an easy-to-use centralized management tool of VMs across the Cloud infrastructure.

The VCC provides a centralized management interface, which is important nowadays considering the lack of automated management services for managing the Cloud infrastructure. For example, if a physical machine fails (e.g., due to hardware failure), the system administrators using the VCC could easily manage it and decide where to migrate the failed VMs. Also, once the failed physical machine is recovered, system administrators need to decide whether to return the VMs back to their original hosting server or leave them at the newly allocated hosting server. Current Cloud infrastructure architects do not allow a VM to automatically migrate/move to a random physical resource in Cloud, as this would cause unwise distribution of resources. The VCC will play a major role in supporting self-managed services, as discussed in subsequent chapters of this book.

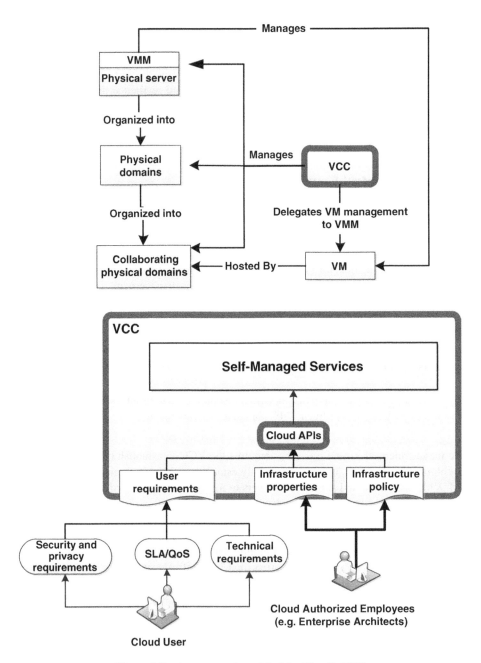

Figure 3.2   A conceptual model of the Cloud's VCC

3.4   Prerequisite Input Data for Management Services

Current implementations of the VCC require excessive human intervention to support the management of the Cloud infrastructure. Clouds are planned to support critical infrastructure and a huge number of users. These necessitate Clouds to be managed automatically with minimal human intervention. The VCC hosts most of the self-managed services and requires trustworthy sources of input data when taking action. This section discusses the types of such data, and subsequent chapters provide detailed discussion of these services and how to establish the required levels of trustworthiness. The main types of input data that would affect decisions made by self-managed services are as follows:

- *Infrastructure properties*. As we discussed earlier, the Cloud physical infrastructure is very well organized and managed by multiple parties, for example enterprise architects, adminis-trators, and security managers. These parties build the infrastructure such that it could host various types of services. Providing different services would necessitate partitioning the infrastructure into groups, each associated with *static properties*. Each set of group-specific static properties is identified based on the properties of its physical resources and their connectivity. The VCC needs a trustworthy source of input data of the static properties and any changes to them.
- *User properties*. A Cloud user interacts with the Cloud provider via a web interface and supplied APIs. This enables users to define their *dynamic properties*. The dynamic properties would cover the following for potential Clouds:
  - *Technical requirements*. These are related to the application requirements when hosted at the Cloud infrastructure, for example DBMS instances that require high availability with no single point of failure, middle-tier web servers that can tolerate failures, and highly computational applications. This enables the Cloud provider to identify the best infrastructural resources that can manage user requirements. A Cloud provider should provide an easy-to-use interface enabling both sophisticated and naive users to use, identify, and provide their requirements.
  - *Service level agreement (SLA)*. This reports the agreement between a Cloud provider and customers. The agreement contains the criteria of service management, for example quality control measures, legal constraints, and operational requirements. The SLA, for example, defines system availability, reliability, scalability (in upper/lower bound limits), and performance metrics. Cloud providers should adhere to the SLA agreements when hosting and managing the services of their customers.
  - *User-centric security and privacy requirements*. Examples include: users needing strin-gent assurance that their data is not being abused or leaked; users needing to be assured that Cloud providers properly isolate their VMs, especially when the VMs share the same physical platforms (i.e., the problem of multi-tenant architecture [5]); users needing to identify the location of the distribution and processing of their data (which could be for legal reasons). Current Cloud providers have full control over all hosted services in their infrastructure; that is, the Cloud provider controls who can access VMs (e.g., internal Cloud employees and contractors) and where user data could be hosted (e.g., server type and location). Users have very limited control over the deployment of their services, no control over the exact location of the provided services, and no option but to trust Cloud providers to uphold the guarantees provided in their SLAs.

- *Infrastructure policy.* Policies are defined by Cloud authorized employees to control the behavior of self-managed services when operating the Cloud environment.
- *Changes and incidents.* These represent changes in user properties (e.g., security/privacy settings), infrastructure properties (e.g., component reliability, component distribution across the infrastructure, redundancy type), infrastructure policy, and other changes (e.g., increase/decrease of requests, failure of a component, network failure).

3.5   Management of User Requirements

The requirements of complex services, especially those associated with critical and/or highly resilient applications, at the time of writing require system analysts to identify them and system architects to create an application-specific architecture. That is, such a process necessitates the involvement of human elements. Human elements are also needed to exchange the architecture, negotiate SLA, implement and deploy the architecture, and maintain the deployed architecture at the Cloud provider.

In contrast, the requirements of simple applications are easier to identify. Cloud providers have the tools which help them to automatically collect, process, and manage simple user requirements, as we discuss next. Cloud computing by definition should support minimal human intervention [6]; therefore, automating the process of identifying, collecting, and managing user requirements, including complex services, is an important subject for the wider adoption of Clouds. This section covers the details of managing user requirements in the context of Cloud computing.

3.5.1   Requirement Management Workflow

This subsection outlines the typical process which is followed by Clouds to collect and manage user requirements. The following subsection discusses some of the main challenges in this process. The following steps outline the workflow that VCC follows when managing user requirements (illustrated in Figure 3.3).

 I. Users would need to identify their application requirements.[4] This is a complex process, which we do not cover in this book. Users then need to decide whether the requirements could be managed automatically or require human elements. Such a decision is not complicated, as the VCC can only collect and manage very basic requirements (at the time of writing), for example the size of VMs and the flavor of the VMI (Virtual Machine Image).

 II. If the identified user requirements could be managed automatically, users would then connect to a selected Cloud provider typically using a web page or via a set of APIs running on the VCC. They then submit their identified requirements and agree on an SLA governing how the Cloud should manage the requirements.

 III. Once the user submits the requirements they would be validated using a software agent running at the VCC.

---

[4] We mean by 'requirements' the deployment and operational requirements and not the application design requirements.

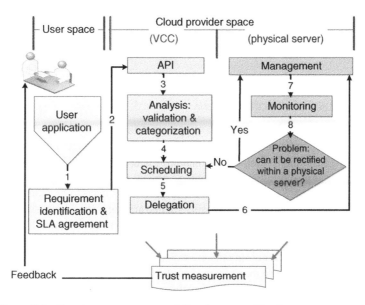

Figure 3.3   Requirement process workflow in current Cloud management tools

IV. If the validation succeeds, the agent categorizes user requirements. As indicated earlier, because of technological limitations current VCC tools only consider limited infrastructure properties and support basic user requirements. As a result, the categorization process is currently related only to individual physical server-related criteria (e.g., CPU speed, size of memory, and available storage space).

V. Next, the agent starts a scheduling process. This process is currently very basic and focuses mainly on matching the categorized user requirements with infrastructure properties. Current implementations of VCCs cover only a few infrastructure properties, which are limited to the capabilities of individual physical servers, such as number of CPUs, memory, etc. Matching user requirements with infrastructure properties would result in a set of physical servers that could host user services. A physical server is then selected from the set of servers to host user services by following a scheduling algorithm. The final result would be stored in a structured database, whose layout we cover in Chapter 12.

VI. Once a physical server is allocated, a delegation process starts. This process sends an 'instantiate VM' message to a management process running at the selected physical server. The message would include the VMI, the requested VM flavor, and other related user requirements.

VII. After the management process receives the message it instantiates a VM and manages it based on the received requirements. Whenever a physical server fails, the VCC would take control and decide where to move the VMs which are hosted at the failed server.

VIII. In principle, users should be provided with tools enabling them to assess the trustworthiness of the process which manages and enforces their requirements. OpenStack [7] has recently introduced a framework for trust attestation which has not yet been implemented.

Having identified the main process workflow for managing user requirements, the next subsection discusses the main problems in this process. It then suggests a list of requirements to address the identified problems.

3.5.2   Challenges and Requirements

As discussed in Chapter 2, Clouds have two types of data: management and application data. We consider user requirements to be part of the management data, as user requirements eventually control the behavior of the Cloud when managing user resources. Therefore, protecting and carefully managing user requirements, which are crucial for establishing trustworthy Clouds, are the responsibility of Cloud providers. User requirements need to be managed throughout the Cloud environment; that is, when requirements are submitted, processed, stored, distributed, and delegated to other entities within the Cloud infrastructure. User requirement protection mechanisms need to provide data integrity, availability, confidentiality, and origin authentication wherever the requirements are transferred and used. Exploiting vulnerabilities in the way the requirements are managed and protected could cause major consequences. For example, an attacker might maliciously update the resilience level of an application, change the hosting location of a VM, and/or change the security properties of an application by attacking the user requirements without the need to directly attack the hosted applications. We now briefly discuss the problems which are associated with the way user requirements are currently managed.

I. Users should not be expected to have sufficient skills to understand the full requirements of their applications, and the way to express and quantify them in technical terms. For example, quantifying the reliability and resilience measures requires high technical skills which are beyond the capabilities of most individuals. In addition, users should not be expected to understand the consequences when changing the values of requirements of their applications. Addressing this requires an easy-to-use and adaptive web interface which helps users to identify their requirements [8]. Such an interface should be intelligent enough to help users understand the consequences of their decisions on the overall architecture of their hosted applications. It could also be designed by following a challenge–response method. That is, a web interface could be designed to dynamically adjust its look-and-feel and its output data based on users' technical abilities and the sophistication of their requirements. The web interface should also ensure that users' requirements do not violate legal requirements.

II. SLAs are complex for normal users to understand [9]. A key requirement in the SLA is that users need to understand whether the complex terms and conditions provided in the SLA are sufficient to assure them that Cloud providers would maintain their requirements. This could possibly be addressed by introducing a trustworthy tool that manages and generates SLAs. The tool should be trusted by both Cloud providers and users. The SLA should, as a result of using the tool, be easy to read by a 'non-technical person,' and should provide the assurance that Cloud would maintain the agreed user requirements.

III. The validation process is critical and complex, especially for sophisticated requirements. For example, it needs to check that the requirements do not contradict each other and can be reflected in the Cloud infrastructure. The validation process does not necessarily need to be implemented only at the Cloud provider side, it could also be part of the adaptive web-interface design (discussed above). This is to ensure that those user requirements (prior submission to the Cloud provider) get validated.

IV. The scheduling process is one of the most important functions of the VCC. Current implementations of schedulers do not consider complex user requirements and also do not consider the complex infrastructure properties. Therefore, we require a trustworthy scheduler that can consider such complexities and scale well with changes in the requirements and properties [10].

V. The scheduler requires trusted and categorized inputs of user requirements and infrastructure properties. Therefore, a process is required which could understand the scope of user requirements in the context of Cloud ontology. Such a process, for example, could help the scheduler to understand how to meet user requirements on the distributed elements of the Cloud infrastructure.

VI. The management of the grouping of entities in the Cloud infrastructure (i.e., domains and collaborating domains) is not yet implemented as part of current VCCs and is currently managed by human intervention.

VII. Current VCCs implicitly assume that Cloud infrastructures are trusted, and as a result user requirements are only protected whilst being transferred between users and the software agent running the VCCs; however, once submitted they are no longer protected inside the environment of Cloud, for example they are transferred in the clear between different entities of the VCC and across related entities of the Cloud, stored unprotected in the VCC database, and when delegating the management of user services to physical servers there is no verification of either the execution environment of the servers whilst processing user requirements or the identity of the physical servers. This requires establishing trust amongst the distributed elements of the Cloud infrastructure, including federated Clouds [11–14]. It also requires establishing federated identity management within Clouds [15].

VIII. The management of user requirements is centralized in such a way that the VCC has to take all decisions for all problems that expand outside a single physical resource boundary. Such centralized management is a single point of failure (see step 8 in Figure 3.3). This, in addition, causes delays in critical operation management, for example service recovery on failure requires additional overheads when communicating with the VCC. Addressing this would require the VCC to be distributed. In addition, it requires the VCC to properly and effectively delegate the management of resources to their domains such that the VCC would only consider highly critical cases while domains should manage the resources which are related to their boundaries.

IX. Users have no way of getting the assurance that their requirements will be managed properly throughout the infrastructure. Therefore, a mechanism is required to enable users to get such an assurance without the need to go into the details of the complex Cloud infrastructure.

Addressing the above points is a complex process which is important for establishing trustworthy Clouds. The remaining parts of this chapter focus on understanding the wider range of user requirements, their generic types, and how their management could be delegated across the distributed elements of the Cloud infrastructure. This partially covers points V and VI. Subsequent chapters of this book discuss points IV, VII, and IX. Other points are still under active research.

3.5.3   Categories and Delegation of User Requirements

This subsection identifies and discusses the main categories of user requirements. It then outlines the management of the delegation process of user requirements. The subsequent subsection clarifies the proposed method on resilience requirements. User requirements could be categorized as follows:

- *Application layer-related user requirements* (or simply application requirements). Application requirements are directly related to an application or a set of applications and would need to be enforced at the application layer. An example of an application requirement includes the measures which control the dependencies between resources of a single and multiple applications, such as whether to host an application resource on a single or multiple VM; the physical distance between the application resources when hosted at different VMs; and the physical distance between dependent applications. In addition to enforcing application requirements at the application layer, they might also need to be enforced at other layers. For example, meeting application availability, reliability, scalability, and resilience might involve collaboration amongst all Cloud layers. The application requirements could also include Cloud services, such as secure logging mechanism as a service and key management as a service.
- *Virtual layer-related user requirements* (or simply virtual requirements). As in the case of application requirements, user requirements could be directly or indirectly categorized as virtual layer-related requirements. An example of directly related virtual requirements are the horizontal and vertical scalability measures. An example of an indirect virtual requirement is the dependency between VMs which is reflected in the dependency between application components, for example a set of VMs needing to be within physical proximity for performance and network congestion reasons, a set of VMs needing to be geographically distant to better serve geographical distribution and for higher resilience reasons, and a set of VMs needing to be hosted at different physical servers but within physical proximity for resilience, performance, and security reasons.
- *Physical layer-related user requirements* (or simply physical requirements). Some user requirements could be reflected directly at this layer, such as geographical location restrictions, physical resources isolation, and binary trust measurements of physical resources. Other user requirements could be related indirectly to the physical layer, such as reliability, availability, resilience, adaptability, and scalability criteria, which would need to be maintained across different Cloud layers including the physical layer. These are demonstrated for resilience requirements in the next subsection.

User requirements are enormous and their management could be delegated at different layers in the Cloud and managed by different elements within the Cloud infrastructure, as pointed out earlier. Careful categorization of user requirements would simplify their management and help in the direction of providing enforcement assurance measures. Some user requirements could be satisfied at multiple layers; however, satisfying a user requirement at one layer could be violated by another layer. The suggested approach, as a result, is to start the delegation from bottom-up, that is to start the delegation at the physical layer, then the virtual layer, and finally the application layer. In other words, the application layer should only manage the requirements that cannot be managed by the physical and/or virtual layers, the virtual layer should attempt to serve the requirements that cannot be managed at the physical layer, and the physical layer should attempt to address as many requirements as practically possible and financially feasible. This would provide a more stringent control and higher levels of security and assurances than a random or a top-down approach. For example, physical layer vulnerabilities could have major consequences on other layers even if the other layers are well protected; however, application layer vulnerabilities might not necessarily affect virtual and physical layers.

3.5.4   Illustrative Example

This section clarifies the bottom-up approach presented in the previous subsection by discussing the management of the resilience requirement. There is a misconception that resilience is only related to physical resources; however, in this subsection we illustrate that resilience does not necessarily require highly resilient physical resources.

Resilience has different levels of granularity which affect the design criteria of the grouping within the various Cloud layers. The level of resilience is the key to know how to delegate its management across the elements of Cloud. We next discuss a possible way in which resilience could be managed across the layers of the Cloud infrastructure.

1. Resilience could be delegated to the physical layer and is reflected as a physical requirement. Users would need to select their desired level of resilience by considering the nature of the hosted application, as follows:

   (a) The first is for critical applications. In this case services must always be available within and across federated Clouds and on emergency the hosted services should immediately be available for customer usage without any downtime. This case requires an active–active configuration within and across federated Clouds (that is, downtime is not tolerable).

   (b) The second is for highly resilient but less critical applications. In this case services should be available on emergencies within a single Cloud, however, high reliance across federated Clouds is optional. That is, if the federated Clouds option is requested and the primary Cloud provider fails, then services should start and be available within an agreed timeframe (typically minutes) at an alternative Cloud provider. This case requires an active–active configuration within a single Cloud, and an active–passive configuration across federated Clouds. That is, short downtime within federated Clouds is tolerable; however, downtime is not tolerable within a single Cloud.

   (c) The last category could be a default option for all other types of application. In this case, services should be accessible 'most' of the time. This case means downtime is tolerable within or across federated Clouds. However, the length of downtime must be quantified and agreed in advance. This category could be managed completely by other layers without involving the physical layer, as discussed next. This is the only case in which the physical resilience requirement could be eliminated completely, but delegated to other layers based on the agreed downtime.

   The first category is the most technically complicated and the most expensive to implement; the second category is still expensive to implement but is less complicated than the first; the last category is relatively the simplest and the least expensive option. The last category is important for overall Cloud services continuity in the long run. These services are associated with many challenges such as security, privacy, performance, and data management. Users will need to define their preferred option when joining a Cloud provider. In addition, users should be provided with the option to decide if they need their hosted applications to be supported by federated Clouds, and they need to decide on the specific category of the above services and the length of time a service should be recovered if ever failed (within and across federated Clouds). Cloud providers will have different pricing models for the different levels of service.

2. Resilience could be reflected as virtual requirements with or without physical requirements (based on the requested level of service and the application nature). Specifically, the first two types of application discussed in points 1a and 1b should (in addition to the physical layer) be managed by the virtual layer; however, point 1c does not necessarily need a physical requirement as virtual layer resilience should be sufficient to satisfy its requested level of service. For example, higher resilience of an application resource would definitely require horizontal scalability of VMs (see Chapter 5). In this case, the VMs should be hosted at different physical servers. A virtual resource resiliance might not be enough for some kinds of application (e.g., those that require transaction survivability or protection against geographical location failure for highly transactional applications). In such a case, both virtual and physical resilience would be required.

3. Resilience can be an application requirement. High resilience at either the virtual or the physical layer would necessitate distributing the access points of an application across all available resources, as in the case of performing horizontal scalability within a multi-tier application scenario. This would require high resilience at the application layer, for example by incorporating a load-balancing algorithm to distribute the incoming and outgoing application requests across available access points.

The above clarifies using an example bottom-up approach to manage user requirements. That is, it discusses a possible approach for delegating the resilience requirement across Cloud layers.

3.6   Summary

Automating the management process of the Cloud environment is a complex task, and a vital requirement for the wider adoption of Clouds, especially by critical infrastructure. The management process is not only restricted to monitoring resources; it also involves resolving problems, and managing infrastructure properties and user requirements. Automating the process of managing Cloud has great advantages, such as: reducing human cost, helping to address the problems of insiders, helping to quantify the level of trust in Cloud, and increasing the overall efficiency and resilience of Cloud.

Currently available Cloud management processes require excessive human intervention and could only automate the management of basic user requirements. This has been the case since the start of IT (i.e., before the Cloud era). The emergence of the Cloud and its potential features necessitates updating traditional management tools. The lack of automated management tools for enterprise infrastructure is due to different reasons, such as: technological limitations, complexity of implementations, lack of standardization, and lack of published solid cases covering user requirements in this domain (that is, the deployment and operational management requirements).

This chapter has covered the basics of Cloud management. Subsequent chapters will build on the concepts discussed in this chapter and extend them to cover other angles of Cloud management.

3.7 Exercises

Q1. Identify Cloud self-managed services and discuss their role in the future development of Clouds.

Q2. What are the Cloud static and dynamic properties?

Q3. Identify and discuss the main challenges of managing user requirements in the Cloud.

Q4. Which critical infrastructures do not yet adopt the public Cloud model?

Q5. Discuss the importance of protecting Cloud management data.

Q6. Discuss the problem of managing user requirements in the Cloud.

Q7. This chapter presented the bottom-up approach for managing user requirements. If the approach is reversed, to top-down, can you think of the disadvantages of top-down in comparison with bottom-up?

References

[1] VMware. VMware vCenter Server, 2012. http://www.vmware.com/products/vcenter-server/.
[2] Microsoft. Microsoft System Center IT Infrastructure Server Management Solutions, 2010. http://www.microsoft.com/systemcenter/.
[3] OpenSource. OpenStack, 2010. http://www.openstack.org/.
[4] OpenSource. OpenNebula, 2012. http://www.opennebula.org/.
[5] Thomas Ristenpart, Eran Tromer, Hovav Shacham, and Stefan Savage. Hey, you, get off of my cloud: Exploring information leakage in third-party compute clouds. In *Proceedings of the 16th ACM Conference on Computer and Communications Security, CCS '09*, pp. 199–212. ACM: New York, 2009.
[6] Peter Mell and Tim Grance. The NIST Definition of Cloud Computing, 2009.
[7] OpenStack. OpenStack Compute – Administration Manual, 2011. http://docs.openstack.org.
[8] Jiming Liu, Chi Kuen Wong, and Ka Keung Hui. An adaptive user interface based on personalized learning. *IEEE Intelligent Systems*, 18(2):52–57, 2003.
[9] P. Wieder, J.M. Butler, W. Theilmann, and R. Yahyapour. *Service Level Agreements for Cloud Computing*. Springer-Verlag: Berlin, 2011.
[10] Imad M. Abbadi and Anbang Ruan. Towards trustworthy resource scheduling in clouds. *Transactions on Information Forensics & Security*, in press.
[11] Jemal Abawajy. Determining service trustworthiness in intercloud computing environments. In *10th International Symposium on Pervasive Systems, Algorithms, and Networks (ISPAN)*, pp. 784–788, December 2009.
[12] Jemal Abawajy. Establishing trust in hybrid cloud computing environments. In *Proceedings of the IEEE 10th International Conference on Trust, Security and Privacy in Computing and Communications, TRUSTCOM '11*, pp. 118–125. IEEE Computer Society: Washington, DC, 2011.

[13] S.M. Habib, S. Ries, and M. Muhlhauser. Cloud computing landscape and research challenges regarding trust and reputation. In *7th International Conference on Ubiquitous Intelligence Computing and 7th International Conference on Autonomic Trusted Computing (UIC/ATC)*, pp. 410–415, October 2010.

[14] Brian Hay, Kara L. Nance, and Matt Bishop. Storm clouds rising: Security challenges for IaaS cloud computing. In *HICSS*, pp. 1–7. IEEE Computer Society, 2011.

[15] Antonio Celesti, Francesco Tusa, Massimo Villari, and Antonio Puliafito. Security and cloud computing: Intercloud identity management infrastructure. In *Proceedings of the 2010 19th IEEE International Workshops on Enabling Technologies: Infrastructures for Collaborative Enterprises*, WETICE '10, pp. 263–265, IEEE Computer Society: Washington, DC, 2010.

4

Cloud Properties

This chapter identifies and discusses the main properties of a Cloud environment. These are adaptability, scalability, resilience, availability, reliability, security, and privacy. Such properties are vital when comparing different Cloud providers. They are also important for Cloud providers when assessing their infrastructure and for introducing different business models. Importantly, Cloud properties provide measures which help in quantifying Cloud trustworthiness. Assessing the level of trust of a Cloud provider is one of the hot research topics in this area. A foundation of this is quantifying the Cloud properties. This chapter defines and discusses these properties; however, quantifying these properties is not an easy task and lots of work still needs to be done in this domain.

4.1 Introduction

There are, of course, many definitions of the word 'trust.' For example, Diego Gambetta [1] defines trust as follows:

> Trust (or, symmetrically, distrust) is a particular level of the subjective probability with which an agent assesses that another agent or group of agents will perform a particular action, both before he can monitor such action (or independently of his capacity ever to be able to monitor it) and in a context in which it affects his own action.

There are also a number of techniques that enable one party to establish trust with an unknown entity: direct interaction, trust negotiation, reputation, and trust recommendation and propagation. Most of these establish trust based on identity. Trust negotiation, by contrast, establishes trust based on properties. In a Cloud context, as we discuss later, establishing trust should be based on both identities and properties. This section discusses the main properties of the Cloud infrastructure that would need to be considered when assessing its level of trust.

Assessing the level of trust in the Cloud infrastructure is not a trivial task. The main points that would need to be addressed are as follows: What are the properties that the assessor should

Cloud Management and Security, First Edition. Imad M. Abbadi.
© 2014 John Wiley & Sons, Ltd. Published 2014 by John Wiley & Sons, Ltd.
Companion Website: www.wiley.com/go/abbadi_cloud

check? What are the factors that help in assessing the properties? How could such properties be assessed? Who should be involved in the assessment process? How can users establish trust in the assessors?

This chapter addresses the first two questions. Other questions are discussed in subsequent chapters of the book. This chapter is organized as follows. Section 4.2 discusses the adaptability property. Section 4.3 discusses the resilience property. Section 4.4 discusses the scalability property. Section 4.5 discusses the availability property. Section 4.6 discusses the reliability property. Section 4.7 discusses the security and privacy by design requirement. Section 4.8 clarifies the importance of the discussed properties in a simplified business model. We summarize the chapter in Section 4.9 and provide a list of exercises in Section 4.10.

4.2   Adaptability Property

The adaptability property is a statistical number reflecting a Cloud provider's ability to react in a timely and efficient way to infrastructure and application changes and events. Examples of changes and events include the following: adding new components, removing components, changing the location of components, acts of God, an application component failure, etc.

The adaptability property should by default consider security and privacy requirements. The factors affecting the value of the adaptability property are as follows:

- *Adaptability as virtual and application services*. These services, as discussed in further detail in subsequent chapters, perform the management role in a Cloud environment, for example infrastructure and application incidents and change management. The two roles manage changes and incidents within the Cloud infrastructure by liaising with other services and/or internal and external teams to ensure the successful resolve of an incident and successful execution of a change. Examples of such changes and incidents include: hardware failures, change in user requirements, security incidents, and so on.

  The factors that help in assessing the adaptability as virtual and application services include the following: mean time to discover (MTTD) an incident, mean time to invoke (MTTI) an action to remedy the incident, and mean time to recover (MTTR) from an incident. In addition to these factors, the *trust* factor is an important factor that should also be considered.

  Providing adaptability as an automated service that does not require human intervention would result in a much quicker incident discovery time, invocation time, and recovery time. This in turn would reduce the values of MTTD, MTTI, and MTTR because self-services do not require a physical human presence at incidents, do not require coordination amongst multiple team members, and do not require critical human observations.

  In traditional enterprise infrastructures trust is related to operational services which are provided by human beings and assessed based on prior experience in the enterprise. Automated services, in contrast, enable better measurement of trust, as the more mature and tested an adaptability service the higher the value of trust it would have.

- *Tolerance to attack.* The value of this is based on two factors: statistical figures on prior experience of the ability of the infrastructure to mitigate attacks caused by insiders or outsiders; and statistical figures about Cloud provider proactivity, which could be estimated based on the security risk management process.

4.3 Resilience Property

The resilience property is a statistical number reflecting the ability of a system to maintain its features (e.g., serviceability and security) despite a number of subsystem and component failures. A high value of resilience requires a design which uses redundancy to eliminate any single point of failure, together with well-crafted procedures (e.g., a procedure that defines a disaster recovery process). Resilient design helps in achieving higher availability and reliability, as its design approach focuses on tolerating and surviving the inevitable failures rather than trying to reduce them.

The factors affecting the value of the resilience property are as follows:

- *Resilience as virtual and application services.* These services resemble the activities of Cloud internal employees (specifically enterprise architects) when producing the infrastructure architecture to eliminate any single point of failure. Examples of resilience services include the following: if a hardware component fails, the system services should not be affected; if a process fails, the system should provide redundant services that support the failed services; and data should be replicated to protect against physical corruption, failures, and/or security attacks. Factors that help in assessing resilience as virtual and application services include MTTD, MTTI, MTTR, and trust. These follow the same description provided for the adaptability property.
- *Adaptability property.* The higher the value of the adaptability property, the better the system can support resilience as virtual and application services, which in turn increases the value of the resilience property. In other words, the better the management of a Cloud environment the more likely the Cloud is to maintain higher resilience.
- *Tolerance to attack.* This follows the same description provided for the corresponding point at the adaptability property; however, it would be reflected in the resilience domain.
- *Architecture.* The architecture of a system affects the resilience level of the infrastructure and includes the following: redundancy and replication of resources; individual component reliability as provided by the manufacturer; and process management that provides automated scripts and documents (these identify exact procedures on incidents).
- *Feedback of availability as virtual and application services and reliability as virtual and application services.* Availability and reliability are the two main properties that are reflected directly by resilience. The higher the resilience of a system, the higher the availability and reliability to be expected. Availability as virtual and application services and reliability as virtual and application services reflect the on-time work performed by a system to maintain its availability and reliability properties. Therefore, getting statistical figures about these services should indicate the effectiveness of resilience as virtual and application services. This in turn affects the resilience property.

4.4   Scalability Property

The scalability property is a statistical figure reflecting the ability of Clouds to enable a virtual infrastructure to scale resources up and down based on demand. For example, obviously at peak periods the virtual layer should scale resources up, and similarly at off-peak periods the virtual layer should release unneeded resources by scaling down. These should be reflected at the application to support the addition and removal of virtual resources. Also, these should not affect fundamental system properties and should always enforce user requirements, such as security and privacy.

Scalability at a virtual layer can be of two types:

- *Horizontal scalability.* The horizontal scalability relates to the number of instances that would need to be added to or removed from a system to satisfy an increase or decrease in demand.
- *Vertical scalability.* The vertical scalability is about increasing or decreasing the size of instances themselves to maintain an increase or decrease in demand.

In this regard, application layer scalability reacts differently to both types of scalability. For example, horizontal scalability means the application will be replicated at the newly created VMs; vertical scalability means the application needs to take advantage of the additional allocated resources (e.g., increase memory usage, spawn additional child processes).

The factors affecting the value of the scalability property are as follows:

- *Scalability as virtual and application services.* These services resemble the work of Cloud internal employees when designing the infrastructure to scale up and down based on demand. It also matches the work of application architects who design the application to take advantage of any additional resources when scaling up and releasing resources when scaling down. For example, if legitimate requests increase beyond the current allocated virtual resource limits, scaling up is then needed to the limits agreed with the customer. The scalability service should add additional virtual resources to cope with any additional demand, and it should also ensure that the application utilizes the additional virtual resources. As an alternative example, if illegitimate requests increase beyond the allocated virtual resources (as in the case of DoS attack), then the system should attempt to address this, for example by scaling horizontally and isolating the attacked VM [2].

  Factors that affect the assessment of scalability service include the following: MTTI, mean time to procure hardware resources (MTTPHW), mean time to deploy (MTT-Deploy), and trust. MTTI and trust follow the same description provided for the adaptability property. MTTPHW describes the average time required to procure additional hardware resources, and MTT-Deploy represents the average time required to deploy additional hardware resources.
- *Tolerance to attack.* This follows a similar description to that provided for the adaptability property, but it should be reflected in scalability-related issues.

- *Failure statistics*. Represent statistical figures when scalability service fails to scale up or down.
- *Scalability statistics*. These include mean time to scale up (MTTS-Up) and mean time to scale down (MTTS-Down). These describe the average consumed time (based on historical figures) when scaling resources up or down, respectively. MTTS-Up and MTTS-Down include: the time required to re-deploy applications when scaling up (vertically and horizontally); the time required to release resources when scaling down; and the time required for the application to utilize additional resources and release unneeded resources when scaling vertically up/down.

4.5   Availability Property

The availability property is a statistical number reflecting the relative time for which a service provides its intended functions. High levels of availability property are a result of excellent architecture, which considers well-crafted procedures, redundant services, and high service reliability; that is, resilient design.

The factors affecting the value of the availability property are as follows:

- *Resilience property*. As we discussed earlier, resilience has a major impact on availability.
- *Availability as virtual and application services*. These services are in charge of distributing requests coming to an application across all redundant application resources. The distribution should be based on a scheduling algorithm which considers various factors on distributing requests (e.g., load). If a resource is down or it is relatively overloaded, the availability as virtual and application services should stop diverting traffic to that resource immediately, and re-divert traffic to other active resources until the adaptability as virtual and application services fixes the problem or until the overloaded resource returns to normal processing capacity. Factors that affect the assessment of availability as virtual and application services include the following: MTTD, MTTI, MTTR, and trust – with the same description as presented for the adaptability property.
- *Tolerance to attack*. This follows the same description provided for the adaptability property; however, it should be reflected in availability-related issues.
- *Failure statistics*. These include statistical figures based on historical experience, including mean time between failure (MTBF) and incident rate.

4.6   Reliability Property

The reliability property is a statistical figure related to the success with which a service functions. High end-to-end service reliability implies a service always provides correct results and guarantees no data loss. Higher individual component reliability, together with excellent architect and well-defined management processes, helps to support higher resilience. This in turn increases end-to-end service reliability and availability.

The factors which affect the reliability property are as follows:

- *Resilience property*. As we discussed earlier, resilience has a major impact on reliability, which in turn necessitates including the value of the resilience property as a factor affecting the reliability property.
- *Reliability as virtual and application services*. These services ensure that an end-to-end service integrity is maintained (i.e., no data loss and correct service execution). If service integrity is affected in any way and cannot be immediately recovered, reliability as virtual and application services would then notify the availability as virtual and application services to immediately bring a service or part of a service down. This is to ensure that data integrity is always protected. Simultaneously, adaptability as virtual and application services and resilience as virtual and application services should automatically attempt to recover the system and notify system administrators in case a decision cannot be made automatically (e.g., data corruption that requires manual intervention by an expert). Factors that affect the assessment of reliability as virtual and application services include the following: MTTD, MTTI, MTTR, and trust – with the same description as presented for the adaptability property.
- *Tolerance to attack*. This follows the same description as provided for the adaptability property; however, it should be considered in the context of the resilience property.
- *Failure statistics*. These include statistical figures based on historical experience including MTBF and mean time to failure (MTTF).
- *Integrity incidents*. These include incident rates which are statistical figures based on historical experience. They also cover certain historical incidents that have caused data integrity problems, such as physical block corruption and memory problems.

4.7   Security and Privacy Property

In the previous sections of this chapter we discussed some of the main properties which help in assessing and comparing various Cloud infrastructures. In this section we discuss the most important properties which currently limit the wider adoption of Clouds, especially by critical infrastructure. These are the security and privacy properties, which should not be independent of the other properties; they rather are considered as an integrated part within the others. That is to say, quantifying security and privacy should be considered as part of other properties and not independently. If a service is not secure then the value of its corresponding property should go down.

In the previous subsections we identified the elements contributing to each of the identified properties. One of these key elements is the services behind the properties. Establishing trustworthiness in such services and the ability to quantify them are key contributors to assessing the operational trust of Clouds. Trust is directly proportional to security and privacy. Trustworthiness means the service performs its job as expected, which includes but is not limited to considering security and privacy by design when performing any action. In other words, security and privacy need to be integrated from within rather than being an added option. This has been considered as a key element for establishing trust in the Cloud [3–10].

The following are examples of security and privacy by design:

- Resilience as virtual and application services should ensure the presence of strong isolation mechanisms between virtual machines when providing a resilient architecture (to address the multi-tenant architecture problem [11]).
- Resilience as virtual and application services should also consider user security and privacy requirements by design. For example, it must maintain user requirements when relocating virtual resources – the allocation of a new physical resource should always be based on user-defined security and privacy requirements as in the case of geographical location restrictions, discussed in [7, 9]. Also, the newly allocated physical resource should not be less secure than the previous physical resource.
- Scalability as virtual and application services must protect VM integrity and confidentiality on replication. It should also permanently remove data on downscaling. The new virtual resources should be allocated based on user properties on a horizontal scaling.
- Availability as virtual and application services should maintain secure communication channels when distributing load, verify the identity of communicating parties, and communicate securely with other services.

4.8   Business Model

In this section we illustrate the importance of the discussed Cloud properties on a simplified business model case.

Cloud users might not be interested in understanding Cloud properties for the overall Cloud infrastructure. A Cloud user's main concern is to ensure that his applications are hosted and managed considering his expectations. Therefore, users would be interested in understanding the values of individual and overall properties of the environment hosting their applications. Knowing the values of individual properties would help Cloud users to assess Cloud providers based on their needs and then judge which provider best matches their requirements.

A Cloud provider might not necessarily maintain a specific and static value for the discussed properties. It is rather expected from a business angle that a Cloud provider would be more interested in providing customers with different service levels by splitting its infrastructure into groups (see Figure 4.1). Each group is associated with hardware and software properties,

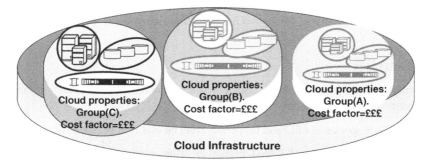

Figure 4.1   A simplified business model illustrating the importance of Cloud properties

which should have a direct impact on the group-specific Cloud properties. In other words, a Cloud provider is most likely to provide different options which are associated with appropriate monetary values. Each option will reflect a guaranteed minimum value of the Cloud properties.

The above business model would fit with the provided abstraction of Cloud layers and resource management within the layers (see Chapter 2). As we discussed in earlier chapters, a Cloud user application will be hosted in a set of domains. Domains that depend on each other join a collaborating domain. A Cloud customer can then request to host his domain or collaborating domains within an infrastructural group that has an expected value of properties. A group that has higher values would be more expensive than a group with lower values.

The variations in service charges in the above business model are due to several reasons, for example: as is well known, different hardware products have different values of statistical reliability – the more reliable a hardware the more expensive it will be; hardware devices have different mechanisms for scaling resources (e.g., some hardware resources permit online scaling of hardware resources – more expensive than offline scaling that requires servers to be shut down before resource scaling); and software tools provided by different manufacturers have different management processes affecting their resilience and reliability.

The above would enable Cloud users to decide on the service level they anticipate. A Cloud provider will then ensure that the Cloud user's set of applications are eventually hosted at resources with specific properties, as agreed in an SLA.

4.9 Summary

Assessing various Cloud providers is an important subject to consider. There are many Cloud providers nowadays. End-users, businesses, and enterprises need a trustworthy and reliable mechanism to help them assess and differentiate amongst the various Cloud providers. Currently, such mechanisms do not exist. This chapter aims to clarify and analyze the properties of the Cloud that could be used to assess different Cloud providers. These properties are as follows: adaptability, scalability, resilience, availability, reliability, security, and privacy. We envisage that Cloud users might be interested in understanding not only the overall Cloud properties, but also the individual properties relevant to their own needs. With suitable trust models and information provision they can judge which provider best fits their requirements.

This chapter does not cover the way to calculate these properties. This is a complex problem to tackle and requires huge collaborative efforts from industry and academia. For example, we do not cover the weights associated with each element of the discussed properties. Making meaningful calculations is non-trivial and is still under research.

4.10 Exercises

Q1. Identify the main properties of a Cloud environment and then discuss their effects on the different Cloud deployment types.

Q2. Discuss the importance of Cloud properties from an end-user perspective.

Q3. Discuss the importance of Cloud properties from a critical enterprise infrastructure perspective.

Q4. How could Cloud properties help in establishing trust in the Cloud?

References

[1] Diego Gambetta. *Trust: Making and Breaking Cooperative Relatioin*. Department of Sociology, University of Oxford, 2000.

[2] Meiko Jensen, Jorg Schwenk, Nils Gruschka, and Luigi Lo Iacono. On technical security issues in cloud computing. *IEEE International Conference on Cloud Computing*, 0:109–116, 2009.

[3] Jamil Abawajy. Determining service trustworthiness in intercloud computing environments. In *10th International Symposium on Pervasive Systems, Algorithms, and Networks (ISPAN)*, pp. 784–788, December 2009.

[4] Imad M. Abbadi. Toward trustworthy clouds' internet scale critical infrastructure. In *ISPEC '11: Proceedings of the 7th Information Security Practice and Experience Conference*, vol. 6672 of *LNCS*, pp. 73–84. Springer-Verlag: Berlin, 2011.

[5] Michael Armbrust, Armando Fox, Rean Griffith, Anthony D. Joseph, Randy H. Katz, Andrew Konwinski *et al.* Above the Clouds: A Berkeley View of Cloud Computing, 2009.

[6] Sören Bleikertz, Matthias Schunter, Christian W. Probst, Dimitrios Pendarakis, and Konrad Eriksson. Security audits of multi-tier virtual infrastructures in public infrastructure clouds. In *Proceedings of the 2010 ACM Workshop on Cloud Computing Security Workshop, CCSW '10*, pp. 93–102. ACM: New York, 2010.

[7] Richard Chow, Philippe Golle, Markus Jakobsson, Elaine Shi, Jessica Staddon, Ryusuke Masuoka, and Jesus Molina. Controlling data in the cloud: Outsourcing computation without outsourcing control. In *Proceedings of the 2009 ACM Workshop on Cloud Computing Security, CCSW '09*, pp. 85–90. ACM: New York, 2009.

[8] S.M. Habib, S. Ries, and M. Muhlhauser. Cloud computing landscape and research challenges regarding trust and reputation. In *Ubiquitous Intelligence Computing and 7th International Conference on Autonomic Trusted Computing (UIC/ATC)*, pp. 410–415, October 2010.

[9] Brian Hay, Kara L. Nance, and Matt Bishop. Storm clouds rising: Security challenges for IaaS cloud computing. In *HICSS*, pp. 1–7. IEEE Computer Society, 2011.

[10] Khaled M. Khan and Qutaibah M. Malluhi. Establishing trust in cloud computing. *IT Professional*.

[11] Thomas Ristenpart, Eran Tromer, Hovav Shacham, and Stefan Savage. Hey, you, get off of my cloud: Exploring information leakage in third-party compute clouds. In *Proceedings of the 16th ACM Conference on Computer and Communications Security, CCS '09*, pp. 199–212. ACM: New York, 2009.

5

Automated Management Services

Automated management services are a key enabler for the wider adoption of Clouds. These services are complex to implement as they represent humans in different domains of expertise and their interactions. This chapter abstracts and analyzes the required Cloud automated management services. We start with the management services of the virtual layer, and then the management services of the application layer. The chapter also discusses the interdependencies amongst these services. Finally, we provide an example scenario using a multi-tier application deployment in Clouds.

5.1 Introduction

Automated management services are a key requirement of the Cloud infrastructure, as identified by NIST (definition in Chapter 1). These services are one of the key areas that distinguish Clouds from traditional enterprise infrastructure [1]. Such services provide Cloud computing with exceptional capabilities and new features. For example, scale per use, hiding the complexity of the infrastructure, automated higher reliability, availability, scalability, dependability, and resilience. These should help in providing trustworthy resilient Cloud computing, and should result in a cost reduction.

Moving the current Cloud infrastructure to the potential trustworthy Cloud infrastructure requires a set of trustworthy self-managed services (also referred to as middleware services). The services should be transparent to Cloud customers and should require minimal human intervention. The implementation of the self-managed services would depend mainly on their location within the layers of Clouds. Self-managed services could be of two types: application layer self-managed services and virtual layer self-managed services. The main objective of this chapter is to identify and analyze these services and their interdependencies. Part Two of this book builds on this chapter and discusses how such services help in establishing trust in Clouds. Also, Part Three presents a partial prototyping of some of these services.

This chapter is organized as follows. Sections 5.2 and 5.3 discuss the virtual layer self-managed services and their interdependencies, respectively. Sections 5.4 and 5.5 discuss the application layer self-managed services and their interdependencies, respectively. Section 5.6 discusses how the automated management services should address security and privacy by

Cloud Management and Security, First Edition. Imad M. Abbadi.
© 2014 John Wiley & Sons, Ltd. Published 2014 by John Wiley & Sons, Ltd.
Companion Website: www.wiley.com/go/abbadi_cloud

design. Section 5.7 illustrates the deployment architecture of a multi-tier application in the Cloud. It then discusses the usage of self-managed services to automatically manage the application in Cloud. Section 5.8 discusses the main challenges and the requirements for providing automated management services. Finally, Section 5.9 summarizes.

5.2 Virtual Layer Self-managed Services

This section provides a set of conceptual models for the virtual layer self-managed services.

5.2.1 Adaptability as a Virtual Service

Figure 5.1 provides a conceptual model of adaptability as a virtual service (ADaaVS), which resembles the role of an infrastructure incident manager. ADaaVS is concerned with adapting virtual resources, which are part of the virtual layer, to *changes* and *incidents*. The changes could be related to user requirements or infrastructure properties. Incidents, on the other hand, have to be related to the virtual layer in some way; for example, there could be a faulty physical resource affecting virtual resources, or an increase in service demand affecting virtual resources. Such changes and incidents should maintain the overall service security and privacy properties as agreed with customers; for example, adding/removing a VM to/from a virtual domain should not compromise the virtual domain security or integrity; and removing physical storage from a physical domain should not reveal content confidentiality.

The primary role of ADaaVS is to perform planning and validation. It then coordinates with other services for execution of the plans. In this context, ADaaVS would always validate the plans with the system architect as a virtual service (SAaaVS, covered next) before taking any

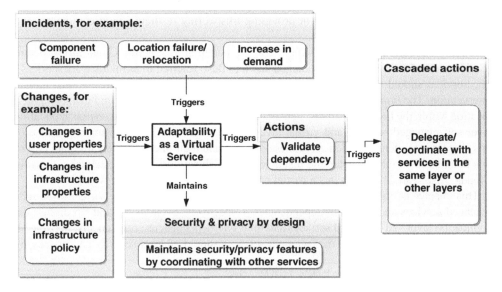

Figure 5.1   ADaaVS

action. This is to ensure that the plan does not have an impact on the system properties. For example, when a group of VMs within a virtual domain requires more resources, ADaaVS would first check if the group is authorized to scale resources up. If so, ADaaVS would then identify all possible resources that might be affected by such scaling, for example scaling the virtual resources of a middle-tier application might also require scaling the backend database resources. Next, ADaaVS would follow the same steps for all identified dependent resources. Once done, ADaaVS would validate the plan with SAaaVS, and then trigger the scalability service to scale up the identified resources.

5.2.2   System Architect as a Virtual Service

Figure 5.2 provides a conceptual model of SAaaVS, which resembles enterprise architect professionals. SAaaVS provides a resilient design and always considers user requirements and infrastructure properties when allocating physical resources to manage virtual resources. SAaaVS is triggered by ADaaVS (e.g., on incidents, when users change their requirements, or when users request a new service).

User requirements could be related to the following properties of the physical layer: resource reliability, redundancy/replication type (e.g., RAID 1+0, RAID 5, dual channel), and the distribution, grouping, and management of resources across the Cloud infrastructure. SAaaVS would also generate and manage well-crafted documents and scripts of process management.

SAaaVS should maintain user security and privacy requirements by design. For example, if a physical domain could not serve a virtual domain for any reason (e.g., network failure), ADaaVS would then check with SAaaVS where to relocate the virtual resources. SAaaVS must ensure that the updated architecture does not compromise user properties.

5.2.3   Resilience as a Virtual Service

Figure 5.2 also provides a conceptual model of resilience as a virtual service (RSaaVS). RSaaVS resembles system administrators who deploy the outcome of SAaaVS at the virtual layer. SAaaVS would provide a resilient design when producing an architecture for a new service request. It would also provide a resilient architecture when updating an exiting architecture based on changes in user requirements or infrastructure properties. RSaaVS communicates with other resources and management tools to deploy the resilient design. It is also in charge of communicating failures of a resource to other services; for example, on resource failure RSaaVS would trigger availability as a virtual service (AVaaVS) to divert traffic to alternative routes and it would trigger ADaaVS to take further action. RSaaVS should maintain security and privacy by design, for example it should consider the hosting of virtual resources at physical domains which are not geographically located within boundaries restricted by the user properties.

5.2.4   Scalability as a Virtual Service

Figure 5.3 provides a conceptual model of scalability as a virtual service (SCaaVS). This service supports the elasticity feature of the Cloud virtual layer by scaling virtual resources up and down when needed. SCaaVS is triggered by ADaaVS when detecting a need to

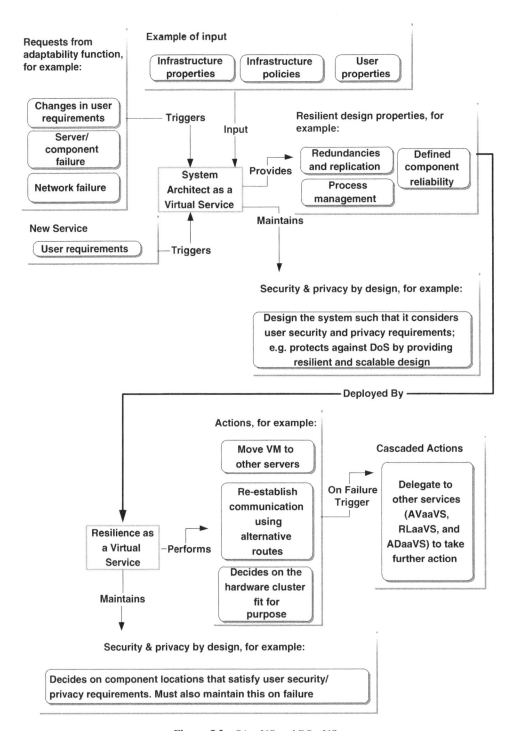

Figure 5.2 SAaaVS and RSaaVS

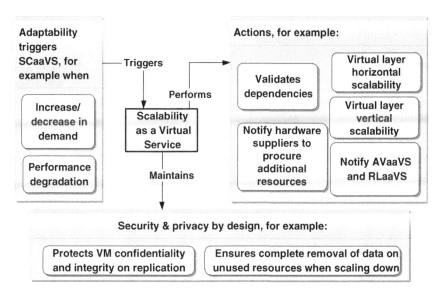

Figure 5.3 SCaaVS

either add or remove virtual resources. In this case, SCaaVS would perform actions. The actions include horizontal scalability by replicating VM resources and/or vertical scalability by expanding VM resources. These actions should always validate user properties before scaling resources and should maintain security and privacy by design. For example: they protect VM integrity and confidentiality on replication; permanently shred data from released resources; and when performing horizontal scaling the allocation of new virtual resources should be based on user properties. SCaaVS should notify AVaaVS and RLaaVS when scaling up/down, as discussed next.

5.2.5   Availability as a Virtual Service

AVaaVS manages the available resources at the virtual layer. Specifically, it is in charge of the following tasks: maintaining communication channels of available virtual services with resources at the application layer and distributing application layer requests evenly across available redundant virtual resources. Availability is supported by a correctly deployed resilient design. The more resilient a system, the higher the availability expected. Figure 5.4 provides a conceptual model of AVaaVS. This figure provides examples of incidents from RSaaVS and changes from SCaaVS which trigger AVaaVS. AVaaVS in turn performs actions based on incidents and changes. The actions also trigger cascaded actions to other services at both the application and virtual layers. For example, if a channel is marked unusable by RSaaVS, AVaaVS would immediately stop diverting traffic to that channel, and re-divert traffic to other active channels until ADaaVS addresses the problem. AVaaVS should always consider security and privacy requirement by design. For example, it should maintain secure communication channels when distributing load, verify the identity of communicating parties, and communicate securely with other services.

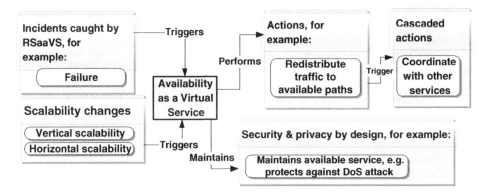

Figure 5.4   AVaaVS

5.2.6   Reliability as a Virtual Service

RLaaVS is in charge of maintaining service reliability at the virtual layer, which is of higher priority than service availability. Most importantly, it ensures that the integrity of the virtual layer management is maintained (i.e., no data loss and correct service execution). If service integrity is affected in any way and cannot be immediately recovered, RLaaVS would then notify AVaaVS to bring the service down by stopping network traffic to the service. This is to ensure that data integrity is always protected. Simultaneously, ADaaVS should automatically attempt to recover the system, and notify system administrators in case a decision cannot be made automatically (e.g., data corruption that requires manual intervention by experts).

Figure 5.5 provides a conceptual model of RLaaVS. This figure provides examples of incidents and changes, which trigger RLaaVS. RLaaVS in turn performs actions and cascaded actions based on incidents and changes. It should also maintain security and privacy by design.

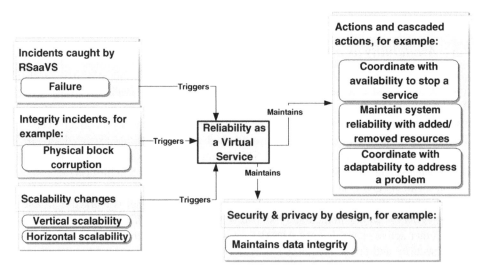

Figure 5.5   RLaaVS

5.3   Virtual Services Interdependency

Figure 5.6 provides a summary of the interactions amongst the virtual layer self-managed services. This figure provides a high-level overview and for clarity is not meant to cover the details in depth.

Humans that manage the Cloud infrastructure need to communicate, exchange messages, and get feedback. Analogously, self-managed services also need to collaborate when managing the infrastructural resources. We discussed this interaction in previous subsections and we now summarize the collaborations between the services. ADaaVS acts as the heart of the virtual layer self-managed services. For example, it intercepts incidents and changes, and then manages them by generating action plans. ADaaVS delegates part of the generated plan to SAaaVS, SCaaVS, AVaaVS, and RLaaVS.

SAaaVS provides a resilient architecture by deciding on component reliability and redundancy. It then triggers RSaaVS to deploy the design. Excellent design results in higher availability and reliability properties. This is indicated in the figure using the *Supports* relation between RSaaVS, AVaaVS, and RLaaVS.

ADaaVS triggers SCaaVS for either vertical scaling and/or horizontal scaling. Once scaling is done, SCaaVS notifies AVaaVS and RLaaVS about the scaling.

RLaaVS is linked with the integrity validation process using the *Must Provide* relation. The outcome of the integrity validation process is fed to RLaaVS. If application integrity is affected, RLaaVS would then send an integrity failure message to both AVaaVS and ADaaVS to take proper action.

5.4   Application Layer Self-managed Services

This section discusses the application layer self-managed services.

5.4.1   Adaptability as an Application Service

Adaptability as an application service (ADaaAS) is the ability to provide timely and efficient support of an application on system changes and events. It resembles the application incident management role that manages changes and incidents at the application layer. This role also coordinates with other management roles at the virtual layer. ADaaAS should always ensure that the overall system properties are preserved (e.g., resilience, availability, and reliability) when taking an action. It should also automatically decide on an action plan and then manage it by coordinating with other services in the same or other layers.

Figure 5.7 provides a conceptual model of the functions of ADaaAS. This figure provides examples of events and changes, which trigger ADaaAS. ADaaAS in turn performs actions based on the events and changes. The actions also trigger cascaded actions to other services in both the application and the virtual layer. ADaaAS follows a set of rules defined by Cloud authorized employees when performing the actions and cascaded actions.

5.4.2   Resilience as an Application Service

Resilience as an application service (RSaaAS) is the ability of a system to maintain an application features (e.g., serviceability and security) despite a number of component failures. High

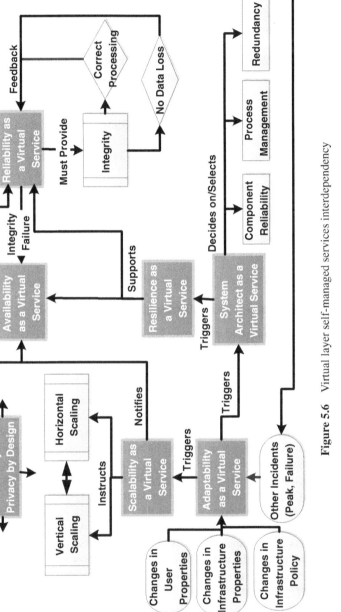

Figure 5.6 Virtual layer self-managed services interdependency

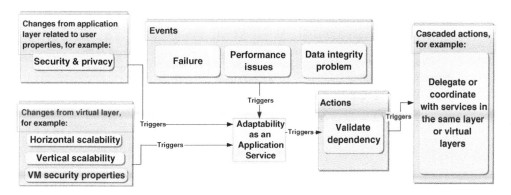

Figure 5.7   ADaaAS

resilience at the application layer can be achieved by providing high resilience at the virtual layer and well-planned procedures, which we discuss in the next subsection. High application layer resilience also requires application redundancy, which can be of the following two types:

- *Active/passive* (also referred to as hot-standby). In this mode two copies of an application need to run in parallel. The first is called active, which receives and responds to service requests. The second is called passive, which maintains an up-to-date copy of the active application but does not process service requests. The passive application can only process requests once its status changes to active when the original active application cannot process requests for any reason, for example network failure, component failure, etc.
- *Active/active* means multiple copies of an application simultaneously process service requests. If any active application copy fails, all requests to that application copy will be diverted to another live application copy.

Resilient design helps in achieving higher availability and end-to-end service reliability, as its design approach focuses on tolerating and surviving the inevitable failures rather than trying to reduce them. RSaaS collaborates with other services to provide an end-to-end resilient Cloud. Figure 5.8 provides a conceptual model for the functions of RSaaS that

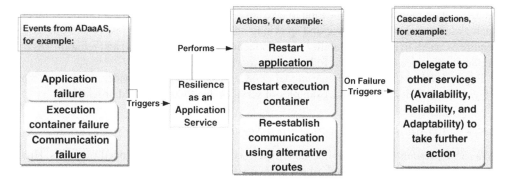

Figure 5.8   RSaaS

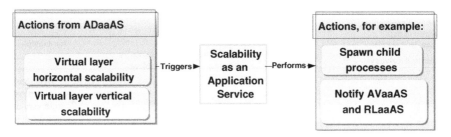

Figure 5.9   SCaaAS

should be provided to maintain the overall end-to-end application resilience. This figure provides examples of a single point of failure which triggers RSaaAS. Once ADaaAS receives notification of a single point of failure event it first manages the event by following a predefined process (e.g., it performs dependencies validation) and then coordinates with RSaaAS. RSaaAS in turn performs actions based on the event. If the actions fail to guarantee a certain level of resilience, a cascaded action plan is then followed. Such actions and cascaded actions follow a set of rules defined by Cloud authorized employees.

5.4.3   Scalability as an Application Service

Scalability as an application service (SCaaAS) is about providing an application with the capabilities to instantaneously[1] and efficiently adapt to the addition and removal of virtual resources. For example, at peak periods the virtual layer scales resources up, and similarly at off-peak periods the virtual layer should release unneeded resources. These should be reflected at the application to support the addition and removal of virtual resources. Also, these should not affect the fundamental system properties and should always represent user requirements (e.g., security and privacy). ADaaVS, upon detecting a need for either adding resources (e.g., peak period) or removing resources, instructs SCaaVS to do this. SCaaVS triggers ADaaAS to adapt to changes at the virtual layer. Finally, ADaaAS triggers SCaaAS to scale the application to adapt to such changes.

Scalability, as discussed earlier, can be of two types: horizontal and vertical. SCaaAS reacts differently to these different types. For example, horizontal scalability of an application requires the application to be replicated at newly created VMs. However, vertical scalability of an application requires the application to take advantage of the additional allocated resources, such as increased memory usage and spawning additional child processes. Also, both cases require SCaaAS to notify availability as an application service (AVaaAS) and reliability as an application service (RLaaAS) to follow an appropriate action plan.

Figure 5.9 provides a conceptual model of SCaaAS. The figure provides the actions that trigger SCaaAS. It also provides examples of services that SCaaAS could provide.

5.4.4   Availability as an Application Service

AVaaAS represents the relative time for which a service provides its intended functions. High levels of availability are the result of excellent resilient design.

---

[1] Existing technologies do not support the instantaneous adaptation.

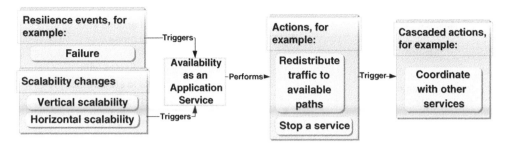

Figure 5.10   AVaaAS

AVaaAS is in charge of distributing requests coming to an application across all redundant application resources based on their current load. If a resource is down or it is relatively overloaded, AVaaAS should immediately stop diverting traffic to that resource, and re-divert traffic to other active resources until ADaaAS manages the problem or until the overloaded resource returns to normal processing capacity.

Figure 5.10 provides a conceptual model of AVaaAS. The figure provides examples of events and changes triggering AVaaAS. The events are triggered by RSaaAS while the changes are triggered by SCaaAS. AVaaAS in turn performs actions based on the events and changes. Such actions could also trigger cascaded actions to other services at both the application and virtual layers.

5.4.5   Reliability as an Application Service

RLaaAS is related to the success with which a service functions [2]. High end-to-end service reliability implies that a service always provides correct results and guarantees no data loss. Higher individual component reliability, together with excellent architect and well-defined management processes, helps in supporting higher resilience. This in turn increases end-to-end service reliability and availability.

RLaaAS is of higher priority than AVaaAS, as RLaaAS ensures that an end-to-end service integrity is maintained (i.e., no data loss and correct service execution). If service integrity is affected in any way and cannot be immediately recovered, then RLaaAS notifies AVaaAS to immediately bring the service or part of the service down. This is to ensure that data integrity is always protected. Simultaneously, ADaaAS should automatically attempt to recover the system and notify system administrators in case a decision cannot be made automatically (e.g., data corruption that requires manual intervention by experts).

Figure 5.11 provides a conceptual model for RLaaAS. The figure provides examples of events from RSaaAS and the virtual layer services, and changes from SCaaAS. Such events and changes trigger RLaaAS. RLaaAS in turn performs actions and cascaded actions based on the events and changes.

5.5   Application Services Interdependency

Figure 5.12 provides a summary of the interaction amongst application layer self-managed services. The figure provides a high-level overview and for clarity is not meant to cover the details in depth. In this figure, ADaaAS acts as the heart of self-managed services. For

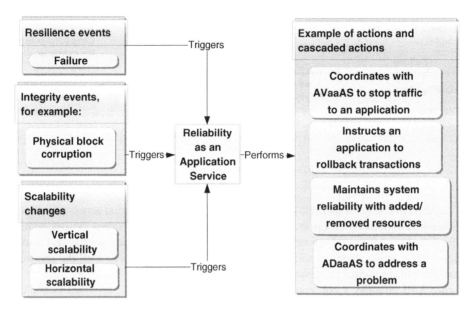

Figure 5.11   RLaaAS

example, it intercepts faults and changes in user requirements, manages these by generating action plans, and delegates action plans to other services. To be in a position to do this, ADaaAS communicates with RSaaAS, SCaaAS, AVaaAS, and RLaaAS.

RSaaAS requires redundant resources, which is represented by the relation *Maintains* on redundancy. An excellent resilient design results in higher availability and reliability properties. This is indicated using the *Supports* relation between RSaaAS, AVaaAS, and RLaaAS.

SCaaAS starts based on triggers which are received from ADaaAS. It instructs to adapt to vertical scaling and/or adapt to horizontal scaling processes. It also notifies AVaaAS and RLaaAS once scaling is done. RLaaAS is linked with the integrity process using the *Must Provide* relation. The outcome of the integrity process is fed to RLaaAS. If the application integrity is affected in any way, RLaaAS would send an integrity failure message to both AVaaAS and ADaaAS.

5.6   Security and Privacy by Design

Security and privacy at the application layer is about ensuring Cloud user security and privacy requirements are maintained by the environment surrounding the application. We do not concern ourselves in this book with the application-related security as we focus on automated services supporting the application and not the application itself. This includes the following examples:

- Protecting Cloud user data whilst in transit (transferred to the Cloud and back to the client, and transferred between Cloud structure components).
- Protecting the data whilst being processed by an application.

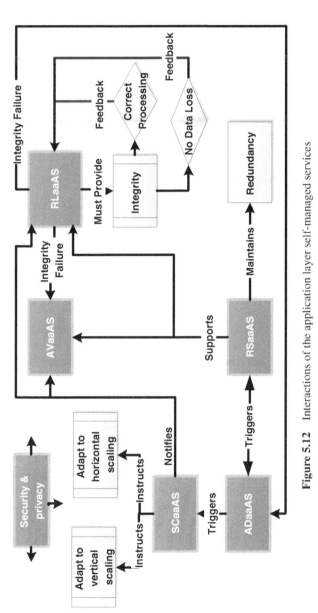

Figure 5.12 Interactions of the application layer self-managed services

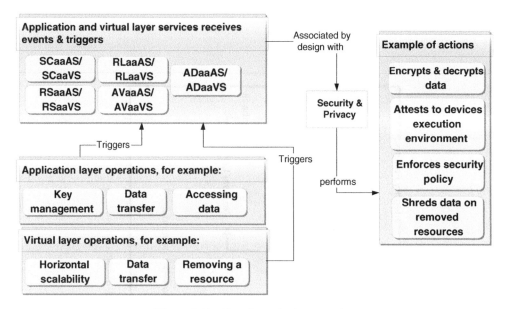

Figure 5.13   Security and privacy service

- Protecting the data when transferred across Cloud services.
- Protecting data whilst in storage.
- Ensuring that the application runs at a certain geographical location as agreed with the customer.
- Ensuring that the application data is stored at a certain geographical location as agreed with the customer.

Security and privacy should be built into all application self-managed services as a default option. Similarly, security and privacy should also be built into virtual self-managed services as a default option. For example, scaling virtual resources should ensure that the selected physical host does not have security holes and has the properties enabling it to serve customer requirements. Also, removing a virtual resource should ensure that the data left are shredded.

Figure 5.13 provides a conceptual model of a *security and privacy* service. This figure provides examples of application and virtual services, which trigger the self-managed services at the application and virtual layer. These services should, by default, ensure that security and privacy are maintained. This is illustrated in the figure using the relation *Associated by design with* between the self-managed services and the security and privacy service. The security and privacy service would perform actions based on the requested service.

5.7   Multi-tier Application Deployment in the Cloud

This section demonstrates the architecture of a typical multi-tier application in the Cloud. It then discusses how the multi-tier application could possibly be managed using the proposed self-managed services.

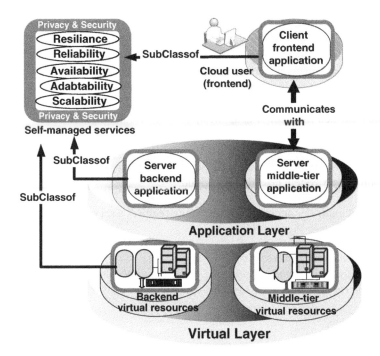

Figure 5.14   A typical multi-tier application architect in the Cloud

5.7.1   Application Architecture

Figure 5.14 illustrates an architecture of a multi-tier application in the Cloud. The application layer would typically be composed of the following components:

- *Server backend application.* In charge of maintaining backend database repository. The database repository runs in an appropriate container (e.g., Oracle DBMS [3], Microsoft SQL Server [4], or Derby [5]). The server backend application would typically be hosted on dedicated physical servers; however, it could also be hosted on a set of dedicated VMs. We refer to the backend application hosting machine as backend VMs.
- *Server middle-tier application.* In charge of running application business logic functions that interact with client frontend applications. The middle-tier application runs in an appropriate container (e.g., Apache/Tomcat [6], Weblogic [7], Oracle Application Server[8]), which would normally be hosted and replicated across a set of VMs. We refer to the middle-tier application hosting machine as middle-tier VMs.

    Middle-teir VMs and backend VMs are usually separate and independent in the production environment for several reasons (e.g., security, resource management, and resilience). These two sets of VMs could be combined and could even be hosted on a single VM for development and testing environment.
- *Client frontend application.* Client application could be a combination of HTML, JavaScript, Java Applets, or even a standalone application that would need to communicate with the Cloud for special purposes (e.g., to upload data on the Cloud for backup purposes or be part

of a supply chain application). Client application could be stored at either the Cloud customer environment or inside middle-tier VMs, based on the application nature. The Cloud customer at run time (possible downloads) runs the client frontend application at the client side.

For example, media organizations usually have editorial systems and online web systems. A media organization could move its online web systems on to the Cloud and keep the editorial applications hosted on their local infrastructure. The organization's editorial employees would then use their local editorial applications when creating and editing stories. The organization's customers, in contrast, would access online web systems from the Cloud. The nature of the client frontend application would typically be a HTML/JavaScript; however, the client frontend application could also be a standalone application which transfers stories into an online web system hosted at the Cloud.

The proposed multi-tier application architecture requires a set of trustworthy self-managed services, as follows (see Figure 5.15).

- *Virtual layer self-managed services.*
- *Application layer self-managed services.* These are conceptually composed of two parts: *server middle-tier self-managed services*, which support the *server middle-tier application*

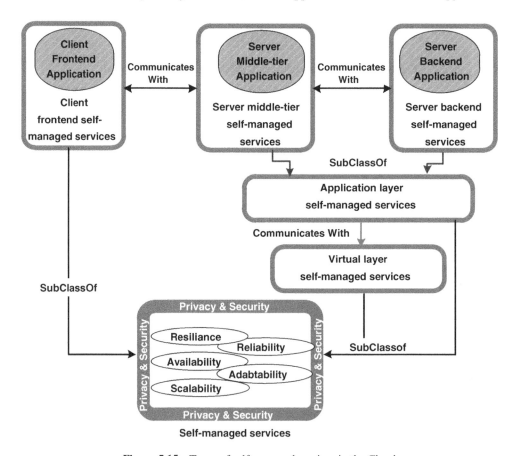

Figure 5.15   Types of self-managed services in the Cloud

and *server backend self-managed services*, which support the *server backend application*. These types of application services would coordinate with each other to support a trustworthy environment for the hosted application. They also need to coordinate with the self-managed services supporting the client frontend application and with the virtual layer self-managed services, as discussed earlier.

- *Client frontend self-managed services*. These should provide transparent management services supporting the client frontend application and its communication with the middle-tier application. This is to provide an end-to-end trustworthy service.

5.7.2 Managed Services Interaction

In this section we illustrate using examples the use of self-managed services to support the management of the discussed multi-tier application architecture. Our approach is based on providing several examples for the interaction between the self-managed services of the client frontend, server middle-tier, and server backend. For simplicity, our examples focus more on the application layer services and less on the virtual layer services.

Client Frontend Self-Managed Services

The client frontend application requires the following self-managed services (we do not discuss issues related to the customer environment's self-managed services as these are not directly related to Clouds). It is important to re-stress at this point that the application is not necessarily a simple HTML, as it could be an interactive application that does part of the application processing at the client application and then communicates with the Cloud for follow-up processes.

- *Adaptability service*. This is in charge of adapting the client frontend application to changes provided by the Cloud provider (i.e., server middle-tier management services), for example changes in service location, degraded performance, and incidents. This would enable an adaptability service at the client side to take appropriate action. Examples of actions include the following:
  - If the middle-tier application location changed, the adaptability service would then send the new location to the adaptability service at the client frontend. The latter would re-establish communication to the new location.
  - If the Cloud application performance was degraded, the client could reduce its requests to the minimal or even do offline processing and then upload the result on the Cloud.
  - The client could temporarily follow an emergency plan on security incidents at the Cloud side.
  These are just a few examples which would be based on the nature of the application.
- *Resilience service*. This is about providing resilient service at the client side when communicating with the Cloud. The service, in this context, mainly attempts to re-establish failed communication with the Cloud (that is, it establishes communication with the server middle-tier services).
- *Reliability service*. This is concerned with maintaining service reliable of the client frontend application when communicating with the Cloud. The service, in this context, ensures reliability when data is transferred from/received by the Cloud, and ensures reliability when data is processed at the client frontend application.

- *Security and privacy*. Integrated with other services to provide security and privacy measures supporting the client frontend application. This includes the following examples:
  - Protecting client's data when retrieved from the Cloud and stored or processed at client environment
  - Protecting data whilst being transferred to/from the Cloud.

Server Middle-tier Self-managed Services

These support server middle-tier applications and require the following self-managed services:

- *Adaptability service*. This is in charge of managing changes and events which could affect the functions of the server middle-tier application. Examples of these include the following:
  - Problems in the Cloud which necessitate relocating the service to another location. In such a case, the adaptability service communicates with the adaptability service at the client frontend to take appropriate action.
  - If the server middle-tier application cannot be restarted because of a hardware-related issue the adaptability service would coordinate with the adaptability service at all other dependent self-managed services.
  - If an application cannot be restarted because of dependency problems, the adaptability service would manage this by finding dependent applications and re-validating their availability.
- *Resilience service*. This covers the following examples:
  - Subject to the nature of the client frontend application, the resilience service re-establishes communications with the client frontend application on failure.
  - It re-establishes communications with the server backend application on failure.
  - It restarts the server middle-tier application on failure.
  - If the application cannot be restarted because of an error (application, environment, or others), the service follows an appropriate procedure based on the error nature (e.g., triggers the adaptability service).
- *Scalability service*. This is mainly concerned with adaptability issues of the server middle-tier application when scaling up/down. It covers the following:
  - Scaling up resources allocated to VMs which host the server middle-tier application. This requires the application to follow a set of processes, for example spawning further child processes.
  - Scaling up by adding a VM which requires the application to follow a different process, for example notifying the availability service to redistribute the incoming load to the newly created VM, and redistributing client sessions considering the new VM.
  - Scaling down by removing additional resources or VMs allocated, in which each requires following a reverse process and then notifying the availability service.
- *Availability service*. This is in charge of distributing the load coming from the client frontend application and server backend application evenly across the redundant resources of the server middle-tier application. If a resource is down, the availability service would immediately stop diverting traffic to that resource and re-divert traffic to other active resources until the adaptability service fixes the problem. Also, when the hosting environment scales up/down the availability service would redistribute the incoming requests based on the nature of the scaling.

- *Reliability service.* This is concerned with maintaining a reliable service for the server middle-tier application when communicating with both the server backend application and the client frontend application. Examples of processes provided by this service include the following:
  - Verifying reliability when data is transferred/received between applications.
  - Verifying reliability whilst data is processed.
- *Security and privacy.* Integrated with other services to provide security and privacy measures supporting the server middle-tier application. This includes the following:
  - Protecting a client's data when retrieved from the client frontend application.
  - Protecting data whilst being processed by the server middle-tier application.
  - Protecting data when transferred to/from the server backend application.
  - Protecting data on storage.
  - Ensuring security and privacy is preserved for all other services (e.g., securing communication baths).

Server Backend Self-managed Services

Required to support the server backend application. These services are the same as those required for the server middle-tier self-managed services. The main difference is that the server backend services do not communicate with the client frontend application. They mainly protect the application which intermediates the communication between the server middle-tier application and the backend storage, where data is eventually stored. This in turn means the implementation of the server backend services would require providing additional functions and security features for managing database instances that interact with the storage.

5.8   Main Challenges and Requirements

This section discusses the challenges involved in providing trustworthy self-managed services. It is beyond the scope of this book to cover the challenges of implementing self-managed services such as data replication.

5.8.1   Challenges

As discussed in Chapter 3, the self-managed services will be managed by the VCC. The VCC communicates with VMMs running at the Cloud infrastructure physical servers to manage the Cloud resources. This process is associated with many challenges, which we categorize in two parts: the first is about providing trustworthy management of the services and the second is for providing trustworthiness in the services.

For the first challenge, the VCC should be trusted to operate as expected; for example, to enforce infrastructure policies and host user virtual resources at physical resources by considering both user requirements and infrastructure properties. Managing self-managed services using the VCC exposes the following issues:

- The VCC, as a central management service, must provide self-managed services without compromising security and data consistency.

- How can the VCC be assured that the VMMs' execution environment is secure, trusted, and reliable to provide timely information about the status of VMs. Also, how can the VCC be assured that the VMMs enforce Cloud user properties.
- How can the VMMs be assured about VCC trustworthiness and about its running execution environment when communicating messages.
- How can the VMM and VCC be assured that their data is stored securely and only accessed when their execution status is trusted.
- The VMM and VCC need assurance about each other's identity.

The second challenge, which is about providing trustworthiness in self-managed services, requires the following:

- As we explained earlier, Cloud's infrastructure is conceptually composed of several intersecting layers. Self-managed services should take into consideration the heterogeneous and complex Cloud layering, and the horizontal and vertical communication channels amongst these layers. Specifically, providing automated self-managed services for a resource requires the following:
  - Understand the relative position of the resource, that is identify the resource's horizontal layer, vertical layer, domain membership, and collaborating domain membership.
  - Understand the infrastructure properties associated with the physical environment hosting a virtual resource (i.e., the properties associated with the physical domain and collaborating domain which host the virtual resource).
  - Understand the user properties associated with a resource in the virtual domain and collaborating domain.
  - Understand how the management of a resource would affect other resources within the same domain.
  - Understand how the management of a resource would affect other resources within the same collaborating domain.
- Policy distribution, coordination, and management across the Cloud entities is a big challenge considering its complex infrastructure.
- The Cloud infrastructure is not hosted at a single data center that is located at a specific location; it is rather the opposite, as it is distributed across distant data centers. This factor has a major impact on decisions being made by self-managed services for several reasons. Examples of such reasons include the following:
  - The distance and the communication medium between distant data centers will have an impact on data transfer speed.
  - Cloud users might have security and privacy concerns about the location of their data.
    Automated services must consider these important factors and other related factors (e.g., data volume, data access mode, etc.) when providing a service.
- Self-managed services must consider federated Clouds and they must also be designed to enforce Cloud provider-related policies when considering a decision to use other Cloud resources. This is important to consider, as federated Clouds could have a major impact on security, privacy, and legislation-related issues. That is, migrating user resources between Cloud providers should ensure user requirements are always enforced.
- Key management is a fundamental issue when discussing Cloud security. This is especially the case as Cloud internal employees are considered potential insiders. Thus, the protection keys of content should not be accessible by Cloud employees.

5.8.2   Requirements

In this section we discuss a set of high-level requirements that could address the problems identified earlier.

A trustworthy management of self-managed services requires the following:

- The VCC and VMMs should attest to each other's execution environment, so that communicating entities can get the assurance of the security and reliability of VMMs and the VCC.
- The VCC and VMMs need to have management agents that are trusted to behave as expected. Such agents are expected to manage the self-managed services across the distributed Cloud infrastructure. The trustworthiness of the agents must be assured to all communicating parties.
- The VCC and VMMs should provide protected storage capabilities.
- The VCC and VMMs should be able to exchange identification certificates with each other in a secure and authentic way.
- The VCC needs to be resilient and scalable to provide distributed and coordinated services.
- The VCC should provide hardware trustworthiness mechanisms to prevent infrastructure single point of failure.

Providing self-managed services requires the following:

- A mechanism to attest to VMMs' trustworthiness to ensure that they would enforce user properties.
- A mechanism to communicate user properties across Cloud-related components, and ensure the properties are not tampered with whilst being transferred, executed, and stored.
- Providing secure information sharing across Cloud components in the same layer and across multiple layers.
- Mitigating the insider threats as discussed next.
- Standardization. Most technologies used in the Cloud are not new; however, the heterogeneous nature of the Cloud requires reconsidering many issues, as in the case of standardization. For example, different software and hardware providers need to provide standard interfaces enabling cross-communication between Cloud components.
- Interoperability. This is an essential requirement to avoid vendor lock problems. In addition, it is the key enabler for collaborative efforts. For example, hypervisor and VMM interoperability enables VMs from different suppliers to work on hypervisors from different manufacturers. This in turn helps in supporting self-managed services.

The Cloud infrastructure must be capable of protecting the integrity, confidentiality, and availability of Cloud critical data from Cloud insiders. This covers all types of data and communication messages, whether directly related to Cloud users or used to manage internal resources; that is, both the application data and the management data discussed in Chapter 2. Part Two of this book presents frameworks which would help in addressing the identified requirements.

5.9   Summary

The complexity of the Cloud infrastructure means a large number of subsystems have to work perfectly together to keep the operation running. In addition, multiple and different groups within and across Cloud layers need to cooperate, exchange critical messages, and coordinate amongst themselves. Current Cloud computing does not provide the full potential of automated self-managed services, and relies on Cloud's employees to support the infrastructure. The complexity of Clouds and their enormous number of users necessitate developing trustworthy self-managed services. Such services are one of the key foundations to address many Cloud problems and for the future success of Clouds.

This chapter has presented a set of conceptual models of self-managed services at the virtual and application layers of the Cloud. The models help in understanding the different types of management services across Cloud layers. They also help in understanding service interdependency. Subsequently, we clarified further the use of the services on a home healthcare application deployed at a Cloud provider. Finally, the chapter discusses the challenges and requirements for establishing trustworthy self-managed services.

5.10 Exercises

Q1. Self-managed services are considered as one of the key distinguishing features of Clouds. Discuss the importance of these services.

Q2. What are the types of self-managed services?

Q3. Discuss the interdependencies between the application layer self-managed services.

Q4. Discuss the interdependencies between the virtual layer self-managed services.

Q5. Discuss how security could be provided using Clouds self-managed services.

References

[1] Keith Jeffery and Burkhard Neidecker-Lutz. The Future of Cloud Computing – Opportunities for European Cloud Computing Beyond 2010.

[2] John D. Musa, Anthony Iannino, and Kazuhira Okumoto. *Software Reliability: Measurement, Prediction, Application*, professional edition. McGraw-Hill: New York, 1990.

[3] Oracle DBMS, 2011. http://www.oracle.com/us/products/database/index.html.

[4] Microsoft Corporation. Microsoft SQL Server, 2008. http://www.microsoft.com/sqlserve.

[5] Derby, 2011. http://db.apache.org/derby/.

[6] Apache, 2011. http://apache.org/.

[7] Weblogic, 2007. http://www.bea.com.

[8] Oracle Application Server, 2010. www.oracle.com/technetwork/middleware/ias/overview/index.html.

Part Two

Cloud Security Fundamentals

6

Background

This chapter helps in understanding the flow of Part Two of this book. It introduces Cloud security, trusted computing, and the general flow of Part Two.

6.1 Topics Flow

This book is about Cloud management and security. We started the first part of this book by discussing the structure of Clouds, their management platforms and attributes. We then discussed the requirements and services that would help in automating the management of Cloud platforms. The second part of this book builds on the first part and assumes readers have carefully grasped the concepts discussed in the first part, especially those relating to Cloud structure. This part focuses on the roadmap and building blocks that are needed to establish the next generation of trustworthy Cloud computing. It also defines the security challenges and the requirements to address these challenges. Establishing trust in Clouds is a big subject in itself. Almost all challenges in Clouds eventually affect the trustworthiness of Clouds, as illustrated in Figure 6.1. For example:

- Addressing the security and privacy challenges of the Cloud is key to establishing trust.
- Addressing the operational management concerns will help in reducing failures, and increasing efficiency and security. These help in enhancing Cloud trustworthiness.
- Addressing the data management concerns will help in increasing data security and reducing outages. These help in enhancing Cloud trustworthiness.

The discussion in this part is centered around issues relating to Cloud security and its effects on trust establishment. Although this part covers the most widely discussed security aspects about Clouds, we do not discuss the security elements topic by topic, neither do we focus on a specific security topic by itself. This is because security is a big subject in itself and we assume readers have a very good knowledge of the foundations of information security. That is, we discuss the process of establishing trustworthy Cloud, which involves many elements

Cloud Management and Security, First Edition. Imad M. Abbadi.
© 2014 John Wiley & Sons, Ltd. Published 2014 by John Wiley & Sons, Ltd.
Companion Website: www.wiley.com/go/abbadi_cloud

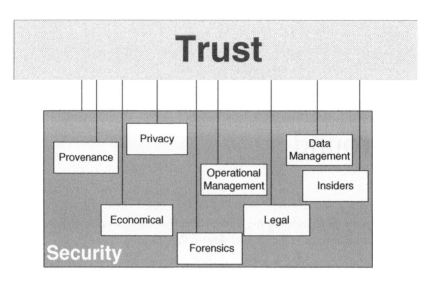

Figure 6.1   Trust components relationship

not limited to information security. In this book we focus more on those elements related to security. This part consists of the following chapters:

- Chapter 6 is the current chapter. It introduces Part Two and highlights its relation to Part One.
- Chapter 7 discusses the challenges related to establishing trustworthy Cloud. The chapter concludes with a set of research directions for establishing trust in Cloud. The remaining chapters in this part extend the identified directions and draw a set of integrated frameworks for establishing the next generation of trustworthy Cloud computing.
- Chapter 8 gives a foundation framework which draws a roadmap to address the question of how to allow users to establish trust in Cloud without the need to get involved in complex technical details.
- Chapter 9 discusses mechanisms for remote attestation in Cloud and addresses the question of how to establish trust in a composition of multiple entities in which the entities could change dynamically.
- Chapter 10 presents a framework for establishing a trustworthy provenance system. This helps in monitoring, verifying, and tracking the operation management of the Cloud infrastructure, that is it helps in the direction of proactive service management, finding the cause of incidents, customer billing assurance, security monitoring (as in the case of lessening the effects of insider threats), security and incident reporting, and tracking both management data and customer data across the infrastructural resources.
- Chapter 11 discusses the insider problem. The insider problem is one of the key areas which restricts the wider adoption of Clouds. It was one of the key security concerns before the Cloud era. The problem is worse with Clouds, as insiders have higher motivation to attack the infrastructure. This is related to the huge number of customers sharing the Cloud infrastructure. A malicious insider can get access to different customer information. This chapter provides a systematic method to identify potential and malicious insiders in a Cloud environment. It also discusses how this would help in having trustworthy Clouds.

6.2   Trusted Computing

This section briefly introduces the fundamental concepts of trusted computing, as it will be very helpful to understand the protocols of the second part of the book.

6.2.1   Introduction

The Trusted Computing Group (TCG[1]) was established in April 2003. It is a not-for-profit organization formed to develop, define, and promote open, vendor-neutral, industry standards for trusted computing building blocks and software interfaces across multiple platforms. TCG is the successor to the Trusted Computing Platform Alliance (TCPA), which was founded in January 1999 by a group of major technology vendors, including AMD, IBM, Intel, HP, Microsoft, and Sun Microsystems (that recently changed name to Oracle).

As discussed in [1] trusted computing platforms provide five main components, which are used to implement TCG functions. These components are: (1) root of trust for measurement (RTM); (2) root of trust for storage (RTS); (3) root of trust for reporting (RTR); (4) root of trust for verification (RTV); and (5) isolation technology. These components are used to implement the TCG functions, which are as follows: (1) an authenticated boot process; (2) platform attestation; (3) protected storage functionality; (4) a secure boot process; (5) software isolation

In the remainder of this section, we start by discussing the core component in TCG specifications (i.e., the trusted platform module). We then briefly discuss the above components and how each function is achieved using these components, as proposed within TCG specifications. Finally, we outline the main criticism raised by researchers on using trusted computing.

6.2.2   Trusted Platform Module

The core component to establishing trust in an IT system based on TCG specifications is the trusted platform module (TPM) [2–4]. TPM is generally implemented as a component which must be physically bound to a platform. A TPM must be completely protected against software attacks and it must be tamper-evident; that is, provide a limited degree of protection against physical attack.

A TPM incorporates various functional components and features, including:

- A cryptographic co-processor that supports the following operations: asymmetric key generation, asymmetric encryption, digital signing capabilities, hashing and random number generation. The asymmetric keys that are generated by a TPM could be either migratable or non-migratable keys. Migratable keys can be transmitted to other TPs if authorized by both a selected trusted authority and the TPM owner. A non-migratable key, in contrast, is bound to the TP that created it, and cannot be cloned.
- An SHA-1 engine.
- Protected storage capabilities. Each TPM has a specific storage root key (SRK), which is securely stored inside the TPM. Once a TPM has been assigned an owner, it generates a new SRK. Other TPM objects (key objects or data objects) are protected using keys that are

---

[1] www.trustedcomputinggroup.org.

ultimately protected by the SRK in a tree hierarchy. Each object protected by a TPM includes a secret 20 bytes of authorization data, which is known as *AuthData*. Proving the knowledge of the value of the AuthData associated with an object grants access to that object.

- Non-volatile and volatile memory.
- Platform configuration registers (PCRs) are special-purpose registers for storing the platform state. Each PCR is a 20-byte inegrity-protected register securely present in a TPM; TCG specifications require that a TPM must contain a minimum of 16 PCR registers.

Each TPM is associated with a statistically unique asymmetric encryption key pair called an endorsement key pair (EK), which is generated either internally or using an external key generator at the time of manufacture. The EK is used only for encryption/decryption purposes. The EK is stored in the TPM at the time of production by the manufacturer. The private decryption endorsement key is known only to the TPM and never revealed outside the TPM. The EK can only be used when assigning TPM ownership.

The TPM is an inexpensive hardware chip, which has protected storage and protected capabilities. In order to reduce TPM costs, the TCG specifications only require the TPM to be used for functions requiring protected storage and capabilities. Functions that do not require protected storage and capabilities can run using the platform main processor and memory space. A significant proportion of all new PCs now incorporate a TPM for implementing trusted computing functionality.

6.2.3   TCG Main Components

We now briefly discuss the five main components of TP.

The Root of Trust for Measurement

The RTM is a computing engine capable of making reliable measurements of TP running components, it is known as an integrity measurement.

> **Definition 6.1   Integrity measurement.** A cryptographic digest or hash of a platform component; that is, a piece of software executing on a platform [5].

In order to ensure that an unlimited number of measurements could be stored in the limited number of PCRs in a TPM, the concept of an integrity metric has been raised, which is defined as follows.

> **Definition 6.2   Integrity metric.** A condensed value of integrity measurements. It is calculated by concatenating a new integrity measurement with the existing content of a PCR, and hashing the resulting string. Following that, the resulting integrity metric replaces the old value of the PCR.

The RTM is controlled by a particular instruction set, which is known as the core root of trust for measurement (CRTM). On a PC, the CRTM may be contained within the BIOS, and is executed by the platform when it is acting as the RTM. The CRTM must be protected against software attack. The CRTM measures the first piece of software to be executed during system boot. Next, it passes the measurement result to the RTS that records the result in the TPM PCRs, and then passes control to the next piece of software to be executed, which has a measurement agent (MA) embedded within it. This MA measures the next piece of software to be executed, passes the result to the RTS that records the result in the TPM PCRs, and passes control to the next piece of software to be executed, and so on. MAs are used to build up a chain of trust in the form of a series of integrity measurements. The results of integrity measurements made by the CRTM and MAs are known as measurement events; these involve two classes of data – measured values, which are representations of embedded data or program code, and measurement digests, which are hashes of the measured values. The measurement digests are stored in the TPM PCRs. The measurement values are stored in the stored measurement log (SML), which is stored outside the TPM.

The Root of Trust for Storage

The RTS is a collection of capabilities, which must be trusted if storage of data in a platform is to be trusted [6]. The RTS uses TPM components to achieve its functions; for example SRK, PCR, cryptographic co-processor. These main functions are as follows:

- Maintaining the integrity measurements made by the RTM by generating the integrity metrics.
- Providing confidentiality and integrity protection to keys and data.

The RTS in the TPM must be immutable, which implies that the replacement or modification of RTS should be under the control of the TPM manufacturer alone.

The RTR enables a TPM to reliably report information about its identity and the current state of the TPM host platform. This is achieved using a set of keys and certificates, which are signed by a variety of third parties that must be trusted if the state of the platform is to be trusted. These certificates are as follows:

- An endorsement credential containing the public EK belonging to a particular TPM. This credential is signed by a trusted platform module entity (TPME), which attests that a particular TPM is genuine. The TPME is likely to be the TPM manufacturer.
- A conformance credential signed by a conformance entity (CE) to attest that the TP design, that is the design of the TPM and other trusted platform building blocks, when integrated into a particular design of platform, meets the TCG specifications.
- A platform credential signed by a platform entity (PE) to attest that a particular platform is an instantiation of a TP design, as described in specified conformance credentials. The PE may be the equipment manufacturer.
- A validation certificate signed by a validation entity (VE) to certify the software components' integrity measurements. The integrity measurements correspond to a correctly functioning platform component (i.e., a piece of software). These validation certificates are used by a challenger wishing to evaluate the state of a challenged TP. The VE is typically the component supplier.

There is still an important set of keys which provide anonymity, and simultaneously attest that a particular platform is genuine. These keys are called the attestation identity keys (AIKs). AIKs (which are signature key pairs) function as aliases for the TP; they are generated by the TPM, and the public part is included in a certificate known as an identity credential. The identity credential asserts that the (public part of the) AIK belongs to a TP with specified properties, without revealing which TP the key belongs to. The first generation of TPM (i.e., V1.1) requires a privacy certification authority (Privacy CA) to certify AIKs, that is signing the identity credential confirms that it belongs to a genuine TP. However, in the second release of TPM specifications (i.e., V1.2) it imposes the direct anonymous attestation (DAA) protocol [7], which allows TPs to authenticate themselves as genuine TPs while preserving their anonymity. The pros and cons for using privacy CA and DAA are discussed later in this chapter.

Before generating an identity credential, the privacy CA verifies a series of signed credentials belonging to the platform, including the endorsement credential, conformance credential, and platform credential. These latter credentials are used to guarantee that an EK belongs to a particular TPM, attest that a TP design meets the TCG specifications, and that a particular platform is an instantiation of a TP design as described in the conformance credentials, respectively. AIKs are used to sign data generated inside the TPM, including the values of PCRs which hold measurements of platform state. AIKs can also be used to sign other keys.

The Root of Trust for Verification

The root of trust for verification (RTV) is defined as a computing engine capable of verifying at least one platform component's integrity measurement against its expected value [8]. Note that the main TCG specifications do not include the RTV, enabling a secure boot process. The RTV was first included in the TCG mobile trusted module [8].

Isolation Technology

Software isolation is about providing a secure environment for protecting executed software confidentiality and integrity. Most proposed mechanisms for providing software isolation focused onVMM technology to provide an isolated secure execution environment and, also, on the use of new process generation provided, for example, by Intel's LaGrande initiative [9].

VM supports multiple operating systems from different vendors that, under the control of VMM, utilize the hardware of a single machine. The hypervisor presents a virtual machine interface to the operating system and arbitrates requests from the operating system in the virtual machines. Thus, the hypervisor can support the illusion of multiple machines, each of which can run a different operating system image. '*Although virtualization offers abstraction from physical hardware and some control over process interaction, there still are problems to be solved. For example, in the x86 architecture, direct memory access (DMA) devices can access arbitrary physical memory locations*' [10].

6.2.4   The TP Main Functions

In this section we discuss the TP main functions.

The Authenticated Boot Process

Establishing trust in a TP starts by having an initial trusted state. Achieving this requires the assurance of the trustworthiness of a platform whilst the platform is starting up during the boot process; that is, an authenticated boot process. The authenticated boot process requires interaction amongst two main TCG components, namely the RTM and the RTS. We now illustrate this in the form of an example for currently available PCs.

Step 1. The CRTM first measures itself and the rest of the BIOS. The result is then passed to the RTS.

Step 2. The RTS condenses the CRTM output and stores it in PCR 0, which is the first one within the 16 PCRs. The measurement values (i.e., prior to condensing) are stored in the SML, which is stored outside the TPM. Control is then transferred to the POST BIOS.

Step 3. The POST BIOS measures the host platform configuration, the option ROM code, and the OS loader. The results are then passed to the RTS.

Step 4. The RTS condenses the POST BIOS output and stores this in PCR 1–5. The measurement values are stored in the SML. The RTS then passes the output to the POST BIOS. Control is then transferred to the OS loader.

Step 5. The OS loader measures the OS.

Step 6. At each stage the result of measuring is passed to the RTS, which condenses and then stores it. Control is then passed again to the next components, exactly as discussed above, until the OS is loaded.

Protected Storage

The protected storage is a fundamental function provided by TPM, which relies mainly on the RTS component to not only ensure data confidentiality and integrity when stored on untrusted devices, but also bind the usage of the protected data to a specific platform when its execution status is in a specific predefined state. In this section we discuss this in detail.

As mentioned before, each TPM has a specific asymmetric encryption key pair known as SRK. The private key is securely generated and stored inside the TPM and is never released outside it. SRK is the root of TPM-protected object hierarchy, which is used to protect all objects underneath it. TPM-protected objects can be of two types: key object or data object, discussed as follows:

- *Key object*. A TPM incorporates a functional component that supports the protected storage capability, which is a cryptographic co-processor. Part of the cryptographic co-processor function generates asymmetric key pairs, where the private part of the key is associated with a data structure containing a set of constraints controlling key usage. For example, forcing the private key to be used only on a specific TPM (i.e., the private key is never exported unencrypted outside the TPM), and forcing the key to be used only when the platform is in a specific predefined state. The asymmetric keys generated by a TPM could be either migratable or non-migratable keys. Migratable keys can be transmitted to other TPs if authorized by both a selected trusted authority and the TPM owner. A non-migratable key, however, is bound to the TP that created it and cannot be cloned.

- *Data object.* Can be either data or symmetric keys that are used to protect bulk data using the platform main processor. The TPM protects a data object's confidentiality by encrypting it using either migratable or non-migratable key at a higher layer of the hierarchy, based on the data object's protection requirements. It also implicitly protects a data object's integrity by associating a 20-byte authorization data (AuthData) with the data object before encryption. When a data object is decrypted, the AuthData is requested and then compared with the recovered value. If the values do not match then the decrypted key object will not be released to the caller. However, if the values match then the value could be released based on the key storage type. More specifically, TP provides sealing functionality, which only enables the decryption of data objects using the same TPM that encrypted it, and only when the host platform is in a predefined state. This is achieved by associating three additional values with the encrypted data object: (1) tpmProof, which is a TPM-specific secret value forcing which TPM can successfully decrypt the data object; (2) digest at creation, which represents the state of the host platform when the data was sealed, enabling the verifier to validate the state of the host platform to ensure that the data was not sealed by rogue software; and (3) digest at release, which specifies the required platform state for releasing the decrypted data object.

Platform Attestation

Establishing trust in a TP is based on the mechanisms used for measuring, storing, reporting, and verifying platform integrity metrics; that is, it relies on the following TP components – RTM, RTS, and RTR. Platform attestation is a method to show to a remote party (the verifier) the status and running environment of a local platform. The remote party needs to trust the attestator to reliably measure and report its configuration. As indicated above, TCG has adopted two different approaches to enabling anonymous attestation: Privacy CA and DAA.

TCG defines integrity management as '*the management of component-information through-out the supply chain to ensure their integrity (tamper-free state) and also to the management of the runtime integrity of the entire Trusted Platform through the correct management of its components, both at load-time and at runtime*' [11]. As described in Section 6.2.3, TP measurements are performed using the RTM, which measures software components running on a TP. The RTS stores these measurements inside TPM shielded locations. Next, the RTR mechanism allows TP measurements to be reliably communicated to an external entity in the form of an integrity report. The integrity report is signed using an AIK private key, and is sent with the appropriate identity credential. This enables a verifier to be sure that an integrity report is bound to a genuine TPM. The term *measurement* is used in various ways, as described below.

- *Loadtime measurements* refer to integrity measurements of TP components made whilst the platform is booting up.
- *Runtime measurements* refer to integrity measurements of TP components that are generated during the operation of the platform, that is after the end of a boot-up sequence.
- *Reference measurements* refer to a collection of digest values of TP components, each of which must be collected from the component manufacturer. This provides an authoritative source of component integrity information, which can be read by a verifier of the state of a TP.

Platform attestation works as follows when a requestor, for example, is seeking a service from a verifier.

Step 1. The requestor sends a request to the verifier.

Step 2. The verifier sends a challenge to the requestor, which includes a nonce, to perform an integrity measurement of the entire platform.

Step 3. The requestor returns a platform integrity report to the verifier (using the RTR). The returned report includes the current platform state, reflected in integrity metrics associated with the sent nonce. This is then signed using the platform AIK private key, and is associated with SML and the appropriate identity credential.

Step 4. The verifier first needs to verify the TPM's signature and the AIK credential.

Step 5. The verifier then needs to verify if it is safe to trust all or part of the software environment running on the platform. This is achieved by getting the reference measurements for each component of the requestor's platform from its manufacturer. The integrity metric provider, that is the hardware manufacturer or software vendor, makes these reference measurements accessible. In this way, the verifier knows both the current integrity status of the component making up the requestor's platform and the source authenticity of those components (as coming from the manufacturer).

Step 6. The verifier needs to identify each component of the requestor's platform and compare the reported measurement against the expected reference measurement value (for each component). If the result is positive, the verifier can provide the requested service.

The Secure Boot Process

The secure boot process mainly uses the RTV TP components, outlined in Section 6.2.3, to extended the authenticated boot process function, discussed in Section 6.2.4, in such a way that the platform state, during the boot process, is reliably captured, compared against measurements indicative of a trustworthy platform state, and then stored. If a difference is found between the measured value and the expected value, then the platform halts the boot process.

Isolated Execution Environment

An isolated execution environment is a fundamental requirement to achieve trust in a TP. Achieving this requires the host platform to provide the following services [5]:

- Whilst a program is being executed it should be protected from external interference, for example, by being accessed using direct memory access.
- Executed programs on the same machine can only communicate via a secure and controlled interprocess communication.
- Executed programs on different machines (hardware or virtual machine) must communicate via a secure communication channel.
- Executed programs must communicate with I/O devices via a a secure communication channel.

6.2.5   Challenges in TCG Specifications

The TCG specifications are large, complex, and based on certain assumptions. In addition to its complexity, building a system that can satisfy such assumptions using today's hardware devices and operating systems is a technical challenge, and is the subject of ongoing research [1, 10]. The following list, based on that given by Gallery and Mitchell [1] and by Sadeghi [10], summarizes these challenges:

- Not only do the TCG specifications use 'hard-coded' cryptographic primitives (e.g., using SHA-1 to compute the integrity metrics) but the used cryptography is also not in accordance with current best practice (e.g., RSA signature and RSA encryption).
- The DAA protocol adapted in TCG specifications is subject to anonymity attack by a malicious DAA issuer, as discussed by Rudolph [12].
- Smyth *et al.* [13] have pointed out a possible privacy vulnerability in the implementation of the DAA protocol; described as corrupt administrator attacks. In this the verifier, with the help of a corrupt DAA issuer, can identify a trusted platform.
- The TCG specifications assume that platform configurations cannot be manipulated after the corresponding hash values have been computed and stored in the TPM's PCRs. Satisfying this assumption requires a secure operating system that is especially designed to consider this requirement. Currently available operating systems can easily be modified, for example by exploiting security bugs.
- The deployment and use of trusted computing based on the TCG specifications requires a fully functioning trusted computing PKI, which is currently unavailable.
- As discussed earlier in this chapter, a verifier can determine the trustworthiness of code from hash values (binary measurements of running code). Such a binary-based attestation mechanism has the following shortcomings:
  - It reveals information about the platform's hardware and software configuration to a verifier.
  - It allows remote parties to exclude certain system configurations.
  - It requires the verifier to know all possible trusted configurations of all platforms.
  - Most importantly, updates in firmware or software, or hardware migrations, result in changed hash values for the updated components. This, in turn, prevents access to data bound to the previous configuration.
  In principle, attestation should only determine whether a system/component configuration has a desired property. Several methods have been proposed to meet this requirement, such as property-based attestation [14–16], anonymous property-based attestation [17], and semantic remote attestation using language-based trusted virtual machines [18].
- The TCG specifications implicitly require the establishment of secure channels between hardware components. TPM chips integrated into currently available devices are connected to the I/O board with an unprotected interface that can be eavesdropped upon and manipulated [19]. Secure channels between hardware components can be established using cryptographic mechanisms supported by an appropriate PKI.
- Currently available trusted platforms come pre-equipped with a TPM chip; however, they do not have isolation technology and CRTM. Therefore, the platform state cannot be reliably measured. This undermines the effect of sealing and platform attestation techniques.

- As discussed in [1], the current generation of TCs has usability and conformance problems. For example, when a TC platform owner enables a TPM he must understand BIOS settings; the TC platform owner is also required to set a TPM owner password; and there are password management issues, as unique passwords may be associated with the TPM owner as well as with data and keys protected by a TPM.

Despite the above problems, TC is on the way to being realized in practise; great support for TC technology is emerging from the open-source community, and from collaborative research projects (e.g., OpenTC and EMSCB). Open-source trusted virtualization layers are being developed by both the Xen and L4 communities [20]. Considering that, and in addition that enterprise infrastructures are more advanced and more managed in comparison with home network environments, 'it seems likely that the technology will succeed first in a corporate setting rather than for home use'.

6.3   Summary

This chapter has clarified the flow of the second part of the book. It briefly discussed trusted computing principles. Part Two of the book uses trusted computing principles for protocol design.

References

[1] Eimear Gallery and Chris J. Mitchell. Trusted computing: Security and applications. *Cryptologia*, 33(3):217–245, 2009.

[2] Trusted Computing Group. *TPM Main, Part 1, Design Principles. Specification version 1.2 Revision 103*, 2007.

[3] Trusted Computing Group. *TPM Main, Part 2, TPM Structures. Specification version 1.2 Revision 103*, 2007.

[4] Trusted Computing Group. *TPM Main, Part 3, Commands. Specification version 1.2 Revision 103*, 2007.

[5] P. England, M. Peinado, and Y. Chen. An overview of NGSCB. In Chris J. Mitchell (ed.), *Trusted Computing*, pp. 115–141. IEE, 2005.

[6] S. Pearson. *Trusted Computing Platforms: TCPA technology in context*. Prentice-Hall: Englewood Cliffs, NJ, 2002.

[7] Ernie Brickell, Jan Camenisch, and Liqun Chen. Direct anonymous attestation. In Vijay Atluri, Birgit Pfitzmann, and Patrick McDaniel (eds), *Proceedings of 11th ACM Conference on Computer and Communications Security*, pp. 132–145. ACM Press: Washington, DC, 2004.

[8] Trusted Computing Group. *The TCG mobile trusted module specification*, 2006.

[9] Intel. Lagrande technology architectural overview. Technical Report 252491-001, Intel Corporation, September 2003.

[10] Ahmad-Reza Sadeghi. Trusted computing – special aspects and challenges. In V. Geffert *et al.* (eds), *SOFSEM*, vol. 4910 of *LNCS*, pp. 98–117. Springer-Verlag: Berlin, 2008.

[11] Trusted Computing Group. *Infrastructure Working Group Architecture, Part II, Integrity Management. Specification version 1.0 Revision 1.0*, 2006.

[12] Carsten Rudolph. Covert identity information in direct anonymous attestation (DAA). In H. Venter, M. Eloff, L. Lebuschagne, J. Eloff, and R. von Solms (eds), *Proceedings of the IFIP TC-11 22nd International Information Security Conference (SEC 2007)*, vol. 232 of *LNCS*, pp. 443–448. Springer-Verlag: Berlin, 2007.

[13] B. Smyth, M. Ryan, and L. Chen. Direct anonymous attestation (DAA): Ensuring privacy with corrupt administrators. In F. Stajano, C. Meadows, S. Capkun, and T. Moore (eds), *Proceedings of Security and Privacy in Ad-hoc and Sensor Networks: 4th European Workshop, ESAS 2007, Cambridge, UK*, vol. 4572 of *LNCS*, pp. 218–231. Springer-Verlag: Berlin, 2007.

[14] Property attestation – scalable and privacy-friendly security assessment of peer computers. Technical Report RZ 3548, IBM Research, May 2004.

[15] Ulrich Kühn, Marcel Selhorst, and Christian Stüble. Realizing property-based attestation and sealing with commonly available hard- and software. In *STC '07: Proceedings of the 2007 ACM Workshop on Scalable Trusted Computing*, pp. 50–57. ACM: New York, 2007.

[16] Ahmad-Reza Sadeghi and Christian Stüble. Property-based attestation for computing platforms: Caring about properties, not mechanisms. In *NSPW '04: Proceedings of the 2004 Workshop on New Security Paradigms*, pp. 67–77. ACM: New York, 2004.

[17] Liqun Chen, Rainer Landfermann, Hans Löhr, Markus Rohe, Ahmad-Reza Sadeghi, and Christian Stüble. A protocol for property-based attestation. In *STC '06: Proceedings of the First ACM Workshop on Scalable Trusted Computing*, pp. 7–16. ACM: New York, 2006.

[18] Vivek Haldar, Deepak Chandra, and Michael Franz. Semantic remote attestation: A virtual machine directed approach to trusted computing. In *VM'04: Proceedings of the 3rd Conference on Virtual Machine Research and Technology Symposium*, pp. 3–3. USENIX Association: Berkeley, CA, 2004.

[19] K. Kursawe, D. Schellekens, and B. Preneel. Analyzing trusted platform communication, In *ECRYPT-CRASH*, 2005.

[20] Paul Barham, Boris Dragovic, Keir Fraser, Steven Hand, Tim Harris, Alex Ho, Rolf Neugebauer, Ian Pratt, and Andrew Warfield. Xen and the art of virtualization. In *SOSP '03: Proceedings of the Nineteenth ACM Symposium on Operating Systems Principles*, pp. 164–177. ACM: New York, 2003.

7

Challenges for Establishing Trust in Clouds

Establishing trustworthy Clouds is the ultimate goal of most current research in the Cloud computing area. This chapter attempts to clarify the huge challenges behind this objective. The chapter concludes with a set of research directions for establishing trust in Clouds. The remaining chapters in this book draw a set of integrated frameworks for the research directions presented in this chapter.

7.1 Introduction

Establishing trust in Clouds is an important subject that is yet to receive adequate attention from both academia and industry [1–4]. The Cloud service model attracts users coming from diverse backgrounds, with a variety of requirements. For example, users will frequently be non-technical end-users or organizations that already have a well-established enterprise infrastructure and might be interested in outsourcing part of their operations into the Cloud. Establishing trust in the Cloud should consider the requirement of different types of users, by providing them with different models. Each model should provide suitable levels of transparency in the context of technical complexities and trust establishment. In addition, trust models are not only beneficial to Cloud users, but also to Cloud providers, collaborating Clouds-of-Clouds, and external auditors. For example, trust assessment helps in the following cases:

- Exposing the components that *must* be trusted or are assumed to be trusted in a Cloud.
- Computing a trust metric for a given Cloud, thus enabling comparison between alternative Cloud providers.
- Cloud providers assessing their own resources' trustworthiness, enabling the Cloud to determine its degree of trustworthiness.
- When Cloud providers collaborate, determining the levels of trust of the resources involved in the collaboration.

Cloud Management and Security, First Edition. Imad M. Abbadi.
© 2014 John Wiley & Sons, Ltd. Published 2014 by John Wiley & Sons, Ltd.
Companion Website: www.wiley.com/go/abbadi_cloud

A trust model also helps users to decide on the resources that are best to host their applications, as users are enabled to determine what could happen to their application when hosted in a specific resource associated with particular properties.

As we discussed earlier, many different deployment scenarios of Clouds have been proposed and adopted, including private, community, and public Cloud models. In community and private Cloud models, users will typically have a relationship of mutual benefit or shared goals with the Cloud service provider; they may also be contractually bound to good behavior. These characteristics give rise to a substantial degree of trust in the Cloud: its architecture is also important, but perhaps less so. By contrast, users of public Clouds are much more reliant upon infrastructure properties in order to establish trust.

This chapter is organized as follows. Section 7.2 clarifies the effects of Cloud dynamics on trust establishment. Section 7.3 summarizes the main challenges and identifies a research agenda for addressing the challenges. Finally, Section 7.4 summarizes the chapter.

7.2   Effects of Cloud Dynamism on Trust Relationships

Chapter 4 discussed the properties of the Cloud which result in its dynamic nature. This section demonstrates the effects of the Cloud dynamic nature on the trust relationships amongst Cloud entities and between users and the Cloud provider. To demonstrate this, we consider the following scenario in which we have two entities, the *trustor* (e.g., an entity representing the Cloud user or a self-managed service) and the *trustee* (i.e., a service point at the Cloud that can be at virtual or application layers), which interact such that the *trustor* establishes 'trust' in the ability of the *trustee* to provide some service S, and to enforce an agreed policy P, when both the *trustor* and the *trustee* have behavior B.

7.2.1   Load Balancing

In terms of our scenario, load balancing means that the *trustee* would share the requests coming from service requestors, including the *trustor*, with other resources, as illustrated in Figure 7.1. The serving resources would be capable of handling similar services. This has an implication

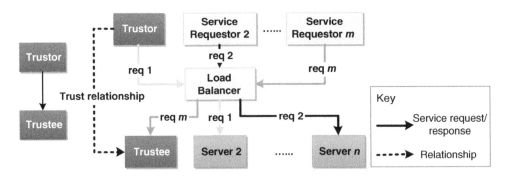

Figure 7.1   Load balancing and trust relationship

that the requests coming from the *trustor* may have to be serviced by a resource other than the *trustee*. Conversely, the *trustee* may have to service requests coming from service requestors other than the *trustor*. The *trustor* maintains a trust relationship initially established with the *trustee*, but the request coming from the *trustor* is not necessarily serviced by the *trustee*.

This scenario demonstrates the need for the *trustor* to update the trust relationship based on the entity assigned to service the request.

7.2.2 Horizontal Scaling

Horizontal scaling for a resource might result in cascaded horizontal scaling at other dependent resources. For example, suppose the *trustee* is a three-tier application within an *application domain*, horizontal scaling of Tier 1 involves introducing more machines at the *virtual domain* that hosts the Tier 1 application component. Increasing the requests coming to Tier 1 is likely to increase the requests going to Tier 2. This in turn might require further horizontal scaling of Tier 2, and so on. As illustrated in Figure 7.2, the *trustee* is no longer three components and this results in a change in the value of B. Horizontal scaling works in conjunction with load balancing, meaning that the effects on trust described above also come into play. More specifically, the request *req* 1 sent from the *trustor* traverses various tiers, some of which are

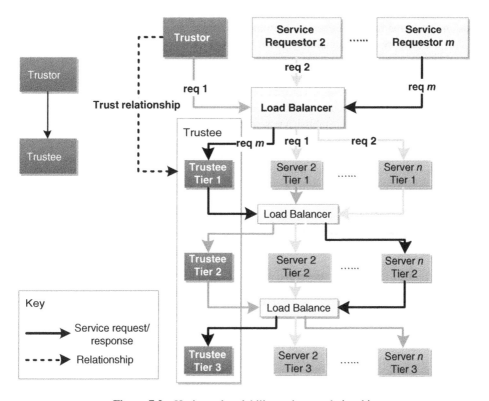

Figure 7.2   Horizontal scalability and trust relationship

not part of the original *trustee*. This invalidates the trust relationship that the *trustor* has in the *trustee*.

7.2.3   Vertical Scaling

For our scenario, vertical scaling implies that the *trustee*'s properties are changed, resulting in a change in the behavior, *B*. In order to maintain a valid trust relationship, the *trustor* would need to re-evaluate the trust decision by checking how the changes to the *trustee*'s properties affect *B*. However, the service provider would like to make the scaling transparent to the *trustor* and therefore a challenge of balancing between transparency of the vertical scaling and maintaining an accurate trust relationship becomes eminent. This problem is due to the fact that if the balance between transparency of the scaling process and trust evaluation is not right, a large window will be created in which the trust relationship will be inaccurate. The balance has to help reduce the size of the window and, if possible, should be coupled with protocols that help to completely eliminate it.

7.2.4   Redundancy

In our scenario, when the *trustee* fails, the system should transparently switch into failure mode in which the redundant server continues to service requests coming from the *trustor*. As shown in Figure 7.3, redundancy has three implications: the *trustee* will have possible trust relationships with other servers; requests coming from the *trustor* may transparently be serviced by a redundant server with which the *trustor* has no direct trust relationship; and the pattern *main server–redundant server* may be repeated multiple times.

In order to maintain an accurate trust relationship, the *trustor* has to invalidate the trust relationship with the *trustee* and establish a new trust relationship with the redundant resource that is now in use. This would need to be done every time the system transitions into failure mode. The challenge here is that the switch into and out of failure mode would normally be transparent to the *trustor*, limiting the chance for the *trustor* to re-evaluate the trust relationship – leaving them to rely on some form of transitive trust.

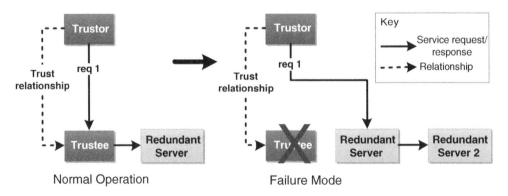

Figure 7.3   Redundancy and trust relationship

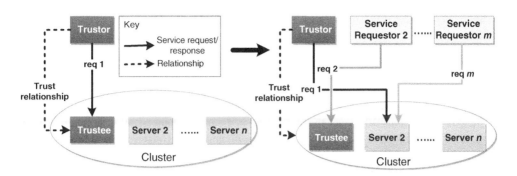

Figure 7.4   Trust relationship for a cluster

7.2.5   Clustering

This is a very important concept in the Cloud environment, which covers the above cases. A cluster could be a group of replicated application resources (i.e., a set of replicated application components within an *application domain*), a group of virtual resources within a *virtual domain*, or a sublayer at a *physical layer*.

In Figure 7.4, we see that initially the *trustor* may establish a trust relationship with a single resource in the cluster. However, when the number of requests increases/decreases or the properties of the cluster change (e.g., full/partial failure of a resource, changing a resource properties, or adding additional resources into the cluster), algorithms that determine the best placement for requests come into play. The main purpose of such algorithms is to ensure that the resources within the cluster are utilized as efficiently as possible and to provide transparency of failures to incoming requests. As a result, no guarantee about the identity of the node that will serve a particular request can be made ahead of time.

A natural solution to this challenge would be to establish trust in an entire cluster. This means the cluster is seen as a single entity by the *trustor*. This, however, brings about different challenges, including: establishing trust in a group of entities and updating the trust relationship as members of the group change state.

7.3   Challenges

In Section 7.2 we demonstrated using several examples of the effects of Cloud management services on 'breaking' trust relationships between a *trustee* and a *trustor*. We noticed that any established relationship could be invalidated at any time as a result of the Cloud dynamic nature. In this section we identify the main challenges in the Cloud environment in the context of trust establishment and outline a research agenda towards addressing these challenges.

- *Compositional chains of trust.* Some entities within the Cloud exist as a composition of multiple entities (e.g., virtual, application, and physical domains). Members of such a grouping may have identical or different chains of trust. However, an entity depending on this grouping should see a single chain of trust representing the trust they have in the grouping. In other words, relying entities will see a single entity, even though that entity will be a grouping representing multiple entities. To address this issue, we set the following research agenda:

Agenda 1. Effective Chain of Trust Functions. Develop functions that can be used to determine the chain of trust from a composition of multiple entities; that is, the effective chain of trust of a physical domain, the effective chain of trust of a group within a virtual domain, the effective chain of trust of an application domain, etc.

- *Trust re-evaluation.* Whenever the state of trusted entities changes, all trust relations with these entities must be re-evaluated and decisions updated based on the changed trust relationships. A challenge here is that there are several scenarios, brought about by the dynamics of the Cloud, which might trigger a re-evaluation of trust decisions. To address this issue, we set the following research agenda:

Agenda 2. Dynamicity Aware Protocols. Develop protocols to effectively support trust establishment and re-evaluation with minimal impact on the desirable properties of the Cloud.

- *Transparency versus trust evaluation.* Many aspects of the Cloud infrastructure require a level of abstraction. For example, the various Cloud service models (IaaS, PaaS, and SaaS) serve as forms of abstraction. Such abstraction requires users at a given level (i.e., IaaS, PaaS, or SaaS) to not deal with internal details of the operation, management, or state of the underlying infrastructure. As a result of such abstractions, several details essential for establishing trust or updating trust may not be made accessible to users, either for security reasons or in order to simplify interactions between users and Cloud providers. In cases where this information is made accessible to users, it might take time for users to re-evaluate trust decisions, and this would have effects on other Cloud properties such as availability. For example, suppose a user at the IaaS level was given information about a detected failure of a resource on which their application is running. The user will then need to consume this information and evaluate the trustworthiness of alternative resources before making a decision. This gets users involved in deep technical details related to the Cloud infrastructure, and the user may take a long time to make a decision, which may lead to service outage. To address these issues, we set out the following research agenda:

Agenda 3. Transparency Strategy. We mean by this finding the right balance between the amount of information made accessible to users against trust evaluation. Preferably, we aim to provide users with mechanisms enabling the user to transparently evaluate trust in the Cloud without the need to get involved in deep technical details about the operation of the Cloud infrastructure.

At the time of writing, there is no single solution to address the above challenges. In subsequent chapters we put forward foundation frameworks to discuss how the above problems could possibly be tackled.

7.4 Summary

The Cloud's desirable properties result in the dynamic nature of the Cloud. This chapter has demonstrated how such a dynamic nature affects trust relationships. The chapter identified the challenges with trust establishment and maintenance. Finally, it defined the research agenda towards addressing these challenges. Subsequent chapters present a set of integrated frameworks clarifying how the identified challenges could be tackled.

7.5   Exercises

Q1. Identify the main parties who would benefit from establishing trust in Clouds.

Q2. What is special about Cloud trustworthiness in comparison with trust in other systems?

Q3. What does *dynamic* mean in a Cloud computing context? What are the main properties that result in Cloud dynamisms?

Q4. Discuss the effects of Cloud dynamic nature on trust establishment. Can you think of other challenges which could result from the dynamic nature of Clouds?

Q5. This chapter discusses the effects of Cloud dynamism on trust establishment. Building on your understanding of this topic, can you think of other challenges of Cloud dynamics?

References

[1] Keith Jeffery and Burkhard Neidecker-Lutz. The Future of Cloud Computing – Opportunities for European Cloud Computing Beyond, 2010.
[2] Michael Armbrust, Armando Fox, Rean Griffith, Anthony D. Joseph, Randy H. Katz, Andrew Konwinski *et al.* Above the Clouds: A Berkeley View of Cloud Computing, 2009.
[3] Sun Microsystems. Take Your Business to a Higher Level, 2009.
[4] Imad M. Abbadi. Toward trustworthy clouds' internet scale critical infrastructure. In *ISPEC '11: Proceedings of the 7th Information Security Practice and Experience Conference*, vol. 6672 of *LNCS*, pp. 73–84. Springer-Verlag: Berlin, 2011.

8

Establishing Trust in Clouds

This chapter presents a framework which draws the roadmap and building blocks to address the last challenge discussed in Chapter 7; that is, the transparency strategy agenda. The framework addresses the question of how to allow users to establish trust in the Cloud without the need to get involved in complex technical details. The chapter identifies a set of requirements, discusses how some of the requirements could be addressed, and leaves other requirements as planned future research work.

8.1 Introduction

There are a number of techniques that enable one party to establish trust in an unknown entity: direct interaction, trust negotiation, reputation, and trust recommendation and propagation. Most of these establish trust based on identity. Trust negotiation, by contrast, establishes trust based on properties. In a Cloud context, establishing trust would be based on both identities and properties [1]. The properties that an attestor requires when establishing trust in Clouds has been discussed in Chapter 4.

This chapter focuses on the provision of a secure and trustworthy environment which assures users that Cloud providers continually enforce their requirements, do not interfere with their application data, and move control of users' application data from the hands of Cloud providers to users. The chapter demonstrates the framework using the IaaS Cloud service and assumes the Cloud user is an organization. The concepts presented throughout this chapter could equally well be applied to other types of Cloud services, as explained in subsequent chapters.

The chapter is organized as follows. Section 8.2 outlines the process followed by an organization when outsourcing the hosting of their application into the Cloud. Section 8.3 identifies the chapter's main objectives and the framework requirements. Section 8.4 identifies device hardware properties. Section 8.5 discusses the dynamic domain concept, and then presents the Cloud framework architecture. Section 8.6 defines and discusses the framework software agents. Section 8.7 discusses the scheme workflow. Section 8.8 analyzes the proposed scheme framework. We summarize the chapter in Section 8.9.

Cloud Management and Security, First Edition. Imad M. Abbadi.
© 2014 John Wiley & Sons, Ltd. Published 2014 by John Wiley & Sons, Ltd.
Companion Website: www.wiley.com/go/abbadi_cloud

8.2   Organization Requirements

Organizations, when outsourcing their applications (or part of their applications), would typically do the following (as clarified in Chapter 3). First, decide on the application that will be outsourced to the Cloud. The application nature, organization policy, and legislation factors would play an important role in such a decision. Then, define the application requirements, which include: technical requirements, service level agreement, user-centric security and privacy requirements. Finally, provide these properties to the Cloud via a set of APIs or a web interface. The APIs are supplied by the Cloud provider, which then creates virtual resources considering the provided user properties. The Cloud provider manages the organizational outsourced resources based on the agreed user properties. In turn, the organization pays the Cloud provider based on a pay-per-use model.

Current Cloud providers have full control over all hosted services in their infrastructure; for example, the Cloud provider controls who can access VMs (e.g., internal Cloud employees, contractors) and where user data can be hosted [2, 3]. Cloud users have very limited control over the deployment of their services, no control over the exact location of the provided services, and no option but to trust the Cloud provider to uphold the guarantees provided in their SLA.

The key requirement for Cloud customers, as a result, is to be provided with a tool enabling them to assess Cloud for meeting their requirements. This tool should not add any extra complexities at the organization side; for example, it should not require organizations to understand the technical details of the Cloud infrastructure. This is a complex requirement to address, which would require a proxy to be setup between Cloud customers and Cloud providers. Both Cloud customers and providers should easily assess the trustworthiness of the proxy. The framework presented in this chapter uses VCC as a proxy, as discussed in the next subsection.

8.3   Framework Requirements

As discussed earlier, the focus of this chapter is on understanding how trust could be established between Cloud customers and the Cloud provider. The approach that we follow uses proxy to establish such trust. The main objective of the proxy is to provide transparent infrastructure management and to remove any extra complexities from the hands of clients. In our discussion we assume the client is an organization that already has an IT infrastructure. The VCC will act as the proxy, and so the organization needs to establish trust in the VCC's abilities to manage and operate its outsourced resources within the Cloud infrastructure based on its requirements. The VCC needs to establish trust with the physical infrastructure, assuring its ability to maintain user requirements. Employees and other resources within the organization need to communicate securely and transparently with the outsourced data at the Cloud without the need to understand infrastructure complexities. Simultaneously, organizations want to ensure that their data is protected at all times. The discussed framework forms the foundation which helps in providing these requirements. This is achieved at two levels, as illustrated in Figure 8.1:

- *Level I.* This is about providing organizations with the capability to control and protect their outsourced resources at the Cloud to the same level of protection and control provided inside

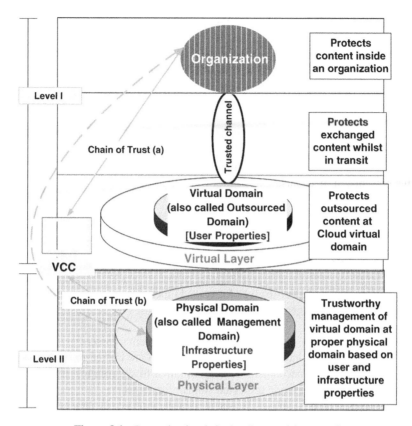

Figure 8.1   Protection levels in the discussed framework

the organization's premises. Level I requires secure and transparent protection measures at three phases:

(a) Protect content inside the organization itself.
(b) Protect content at the Cloud provider side.
(c) Protect content whilst being transferred between the organization's internal resources and the organization's outsourced resources at the Cloud.

- *Level II*. This is related to providing the organization with the capability to attest the trustworthiness of a Cloud provider for managing the organization's outsourced resources. This involves establishing a chain of trust between the organization and the Cloud infrastructural resources (see the dashed arrow in Figure 8.1). One of the key features of Clouds is providing organizations with a transparent infrastructure management. Therefore, such a chain of trust needs to be established at two stages:

(a) A chain of trust between an organization and the VCC.
(b) A chain of trust between the VCC and the Cloud's infrastructural resources.

We now 'informally' discuss the requirements to achieve the above objectives. The requirements can be split into three groups: the first is about the chain of trust (a and b), the second is about Level I, and the third is about mitigating insider threats which affect the discussed requirements (i.e., the chain of trust and Level I). Note that Level II is related to chain of trust (b).

Establishing chain of trust (a), which is part of Level I (see Figure 8.1), requires the organization to attest to the trustworthiness of VCC management agents to manage the organization's resources. Such management, as discussed earlier, should consider both the user and infrastructure attributes. Level I requires the following:

- Data should be protected from within rather than from outside, that is organizations should strongly bind access rights with data.
- The data should not be revealed in the clear except in a trusted execution environment and with the presence of mechanisms to enforce associated access rights.
- Only organizations should be able to access and manage their deployed applications and data running in VMs, that is organizations should have full control of who can access their data.
- The system should provide organizations with superior control of whether to accept the hosting environment properties.

Establishing the chain of trust (b) (see Figure 8.1) includes the provision of a trustworthy Cloud virtual resource management using VCC. This requires the following:

- VCC and VMMs should attest to each other's execution environment.
- VCC and VMMs need to have management agents that are trusted to behave as expected. Such agents should have mechanisms to ascertain their trustworthiness to remote parties.
- VCC and VMMs should provide protected storage functions.
- VCC and VMMs should be able to exchange identification certificates with each other in a secure and authentic way.
- VCC needs to be resilient and scalable to provide distributed and coordinated services.
- VCC should provide hardware trustworthiness mechanisms to prevent infrastructure single point of failure.

In addition to the above, both Level I and Level II involve protecting organizational data from insiders. This requires the following:

- Content must always be encrypted on storage and whenever it leaves a VM; that is content must not leave a VM unprotected either by physical (e.g., copying it to a USB stick) or digital means (e.g., sending it via the Internet). This ensures that only entities having a copy of the content encryption key can access the VM content.
- A content encryption key must be generated automatically away from human observance, not available in the clear, and managed automatically (i.e., distributed and revoked) by a trusted entity.
- Content is revealed only if the environment is verified as trusted and with the presence of a predefined usage policy.
- VM migration/motion between different hardware servers should be carefully controlled and managed, so that VMs are not able to process sensitive data at any time on less secure environments (e.g., insiders might migrate a VM to a less secure hypervisor to indirectly access VM memory space).

- The infrastructure should be highly reactive and resilient (e.g., as in the case of peaks due to insider/outsider attack or fault).
- Employees having access to hypervisors should not be capable of accessing VM space directly from the hypervisor. The hypervisor should also provide full isolation between running VMs.
- Lastly but probably most importantly in a Cloud environment, the above points must consider security of information sharing across VMs, which must collaborate to achieve their mission. Such secure information sharing is still very important but of less concern when organizations own and manage the infrastructure in their internal protected network. For the Cloud, however, the story is different as the Cloud provider network is accessed by many more people who do not have a direct contract with the organization, for example a Cloud provider's system administrator and suppliers.

8.4 Device Properties

A key requirement for addressing the identified challenges is to use devices which are enhanced with trusted computing technology. That is, devices which incorporate a TPM as defined by the TCG specifications [4]. Trusted computing systems are platforms whose state can be remotely tested, and which can be trusted to store security-sensitive data in ways testable by a remote party. The TCG specifications require each TP to include an additional hardware component to establish trust in that platform. This is the TPM, which has protected capabilities; for example protected storage and processing. The entries of TPM PCRs, where integrity measurements are stored, are used in the protected storage mechanism. This is achieved by comparing the current PCR values with the intended PCR values stored with the data object. If the two values are consistent, access is then granted and data is unsealed. Storage and retrieval are carried out by the TPM.

8.5 Framework Architecture

This section presents a framework that forms the foundation for addressing the identified requirements in Section 8.3. The framework uses the dynamic domain concept proposed in [5]. We start by defining the dynamic domain concept, and then discuss the adaptation of such a concept in the framework.

8.5.1 Dynamic Domain Concept

Definition 8.1 A **dynamic domain** represents a group of devices that need to share a pool of content. Each dynamic domain has a unique identifier i_D, a shared unique symmetric key k_D, and a specific PKL_d composed of all devices in the dynamic domain. k_D is shared by all authorized devices in a dynamic domain and is used to protect the dynamic domain content whilst in transit. This key is only available to devices that are a member of the domain. Thus, only such devices can access the pool of content bound to the domain. Each device is required to securely generate for each dynamic domain a symmetric key k_C, which is used to protect the dynamic domain content when stored in the device.

8.5.2   Proposed Architecture

Our framework architecture is composed of the following (see Figure 8.2): *cloud management domain* (MD), *cloud collaborating management domain* (CMD), *organization outsourced domain* (OD), *organization collaborating outsourced domain* (COD), and *organization home domain* (HD). MD, CMD, OD, and COD are hosted at the Cloud provider, while HD is hosted at the organization.

We now map the above domains using the Cloud infrastructure taxonomy concept, as discussed in Chapter 2. A Cloud provider MD and CMD represent a physical domain and collaborating physical domain at the Cloud infrastructure physical layer. An organization OD and COD represent a virtual domain and collaborating virtual domain at the virtual layer. An organization HD includes all devices hosted at an organization, which need to communicate with the OD. We now discuss these entities in the context of the dynamic domain concept.

A management domain is defined as follows:

Definition 8.2   An MD represents a group of devices at the Cloud physical layer. The capabilities of member devices of the MD and their interconnection reflect the overall properties of the MD. Such properties enable the MD to serve the part of user requirements which can only be matched at the physical layer.

An MD has a specific policy defined by the Cloud architect to manage the behavior of MD members when providing services to the OD, when collaborating with other MDs, and during incidents. Such a policy, which is controlled by the VCC, helps in providing Cloud properties (e.g., availability and resilience), which are reflected at virtual resources hosted by the MD. The MD has credentials consisting of a unique identifier i_{md}, a unique symmetric key k_{md}, and a public key list (PKL_{md}). These are defined as follows:

Definition 8.3   The MD identifier i_{md} is a unique number that we use to identify an MD. It is securely generated and protected by the TPM of the VCC.

Definition 8.4   The MD key k_{md} is used to protect the management data that controls the behavior of the MD. k_{md} is a symmetric key that is securely generated and protected by the TPM of the VCC. k_{md} is not available in the clear, it is shared between all member devices of the MD, and it can only be transferred from the VCC to a device when the device joins the MD.

Definition 8.5   The MD's public key list (PKL_{md}) is an MD-specific list that is composed of the public keys of all devices of the MD. Enterprise architects assign devices to each MD by providing each device public key to the VCC in the form of a PKL. The PKL_{md} is securely protected and managed by the VCC.

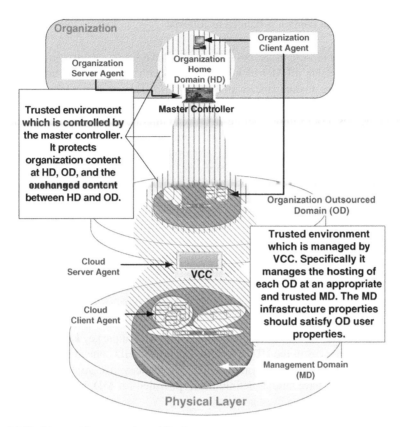

(a) End-to-end framework architecture

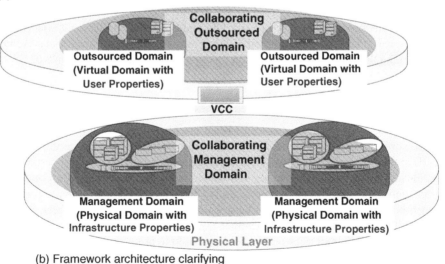

(b) Framework architecture clarifying
collaborations inside the Cloud infrastructure

Figure 8.2 Framework architecture

A collaborating management domain is defined as follows:

Definition 8.6   A CMD represents groups of MDs that have similar infrastructural properties. Such groupings enable all MD members of a CMD to serve as backup for each other, in case of MD failure, maintenance window, or overloaded resources. This operation is controlled by a defined policy at the VCC. Such a policy considers user properties and infrastructure properties. For example, before an OD can migrate from MD_1 to MD_2, both MDs must be within the same CMD and the user requirements must be validated against MD_2 properties, for example users might have restrictions when migrating across different legal jurisdictions, and users might require to migrate CODs for performance reasons when migrating a resource across far-distant data centers.

A home domain is defined as follows:

Definition 8.7   A HD is an organization-specific domain which consists of all organization devices that need access to the OD. HD membership is controlled by an organization-specific master controller, as in Definition 8.8. HD has a specific policy defined by the organization to control the interaction between OD and HD.

As in the case of MD, the HD has a unique identifier i_{hd}, a shared unique key k_{hd}, and a specific PKL_{hd} composed of all devices in the HD. k_{hd} is used to protect HD content that needs to be shared between HD devices. This key is only available to devices that are members of HD. The HD domain credentials have similar definitions as provided for MD, but in this case the HD is managed by the organization master controller.
   A master controller is defined as follows:

Definition 8.8   An MC is an organization-specific trusted device that is in charge of managing organizational domains (i.e., HD, OD, and COD). The master controller enforces organizational policies for domain membership. It runs a trusted server agent that is in charge of performing the master controller's main functions, as explained in Section 8.6.

An outsourced domain is defined as follows:

Definition 8.9   An OD consists of the virtual machines which host organizational outsourced applications at the Cloud. The VCC would establish and manage the hosting of the OD at the MD and CMD based on user properties and infrastructure properties. However, the key management of OD members should be fully controlled by the organization master controller. Each OD has a specific policy defined by the organization to control the interaction amongst its members, and between its members and HD members. As in the case of MD, the OD has a unique identifier i_{od}, a shared unique key k_{od}, and a specific PKL_{od} composed of all devices in the OD. k_{od} is used to protect OD content that needs to be shared between OD devices. This key is only available to devices that are members of the OD. OD credentials have similar definitions as provided for MD, but managed by the organization master controller.

A collaborating outsourced domain is defined as follows:

Definition 8.10   A COD consists of groups of related ODs that share a common policy. For example, the policy could state that COD members should be hosted within physical proximity for performance reasons and COD members should not be hosted in the same MD as in the case of a primary and standby DBMS. Such a policy is defined based on user properties but controlled by the VCC.

8.6   Required Software Agents

The framework requires four types of software agents (as illustrated in Figure 8.2): Cloud server agent, Cloud client agent, organization server agent, and organization client agent. These software agents run on trusted devices that must all have TP properties, as outlined in Section 8.4. The Cloud server agent runs on the VCC; the Cloud client agent runs on a physical member device of the MD; the organization server agent runs in the master controller (see Definition 8.8); the organization client agent runs on member devices of the OD and HD. For convenience, we use the word 'content' to mean either infrastructure management data or organizational data. Organization client/server agents manage organization data, while Cloud client/server agents manage infrastructure management data.

Assumption 8.1   We assume the identified software agents are designed in such a way that they do not reveal domain credentials in the clear, do not transfer domain protection keys to others, and do not transfer sensitive domain content unprotected to others. Although this is a strong assumption recent research shows promise in the direction of satisfying it [6]. TCG-compliant hardware using the sealing mechanism alone is not enough to address such an assumption. Trustvisor [6] moves one step forward and focuses on protecting the content encryption key by utilizing recent developments in processor technology (e.g., Intel TXT). However, this does not protect clear text data once decrypted and more work is required in this direction.

8.6.1   Server Agent Functions

The Cloud and organization server agents have the following shared functions:

- Create and manage domains. This includes the following: securely generating and storing domain protection keys; attesting to the execution environment status of devices whilst being added to the domain and ensuring they are trusted to execute as expected and hence trusted to securely store the domain key and to protect domain content; adding and removing devices to and from a domain by releasing the domain-specific key to devices joining the domain; and backing up and recovering domain-specific credentials.
- Control the exchange of content between different domains.
- Manage policies, and ensure protected content is only accessible to authorized devices. This covers the collaboration between related MDs forming a CMD and related ODs forming a COD.

In addition to the above shared functions, each server agent supports additional functions as follows:

- Organization server agent manages the communication between member devices of the HD and member devices of the OD and COD. It also installs and manages organization client agents at member devices of the HD and OD.
- The Cloud server agent manages the MD and CMD and installs and manages the Cloud client agent at member devices of the MD. Most importantly, the Cloud server agent matches user properties with Cloud infrastructure properties, as associated with the MD resources. This enables the Cloud server agent to manage the hosting of the OD resources on appropriate MD resources by considering both user and infrastructure properties.

8.6.2   Client Agent Functions

The framework has the following two types of client agent:

- Organization client agents are used by devices when interacting with the master controller for joining to the following domains: HD, or OD and COD. Organization client agents are also used to create content and bind it to a domain.
- Cloud client agents have the following functions: they are used by physical devices when interacting with the Cloud server agent for joining an MD and CMD; they manage and enforce MD and CMD policy as defined by the Cloud server agent (e.g., manage hosted virtual resources, manage resilient architecture inside an MD and CMD, etc.).

In addition to the above, both types of client agent are responsible for ensuring that access to their domain content is granted only to applications running on specific devices conditional on having a trusted execution environment. Applications requiring access to content would need to communicate with the device-specific client agent to get access to clear text content. In this case the client agent first verifies the application requesting access is trusted. If so, it decrypts the content and then releases it to the trusted application.

8.6.3   Server Agent Initialization

This section describes the procedure for initializing the organization and Cloud server agents discussed in Section 8.6.1. The main objective of this procedure is to prepare the server agent to implement the scheme framework and manage the domain membership. This includes the following: system administrators install the server agent on its hosting device (i.e., the MC for an organization server agent and the VCC for a Cloud server agent). The server agent installation includes generating a non-migratable key pair (Pr, Pu) to protect domain credentials; the server agent manages security administrator(s) credentials and securely stores them to be used whenever administrator(s) need to authenticate themselves to the server agent.

The first time security administrators run the server agent, it performs the following initialization procedure (as described in Algorithm 8.1). The objective of this algorithm is to initialize the server agent. The server agent executes and sends a request to its hosting device TPM (i.e., to the VCC TPM or MC TPM) to generate a non-migratable key pair, which is used to protect domain credentials. The TPM then generates this key and seals it to be used by the server agent when the hosting device execution status is trusted.

The server agent then needs to ensure that only security administrators can use the server agent. For this the server agent instructs security administrators to provide their authentication credentials (e.g., password/PIN), as described in Algorithm 8.2. The objective of this algorithm is to enrol system administrators in the server agent. The server agent then requests the TPM to store the authentication credentials of the organization security administrators associated with its trusted execution environment state (i.e., the integrity measurement as stored in the TPM PCR) in the hosting device protected storage. By storing data in a protected storage we mean 'sealing data' in TCG terms. The authentication credential is used to authenticate security administrators before using the server agent; see Algorithm 8.3.

Given the definitions and assumptions above, the protocol is described by Algorithms 8.1, 8.2, and 8.3. The objective of the protocol is to install the server agent at the hosting device, which generates the non-migratable key to encrypt the domain credentials and securely store a copy of the security administrator's credentials. The protocol is used every time security administrators want to manage the domain. The following notation is used in the algorithms: M is the software agent; TPM is the TPM on the software agent hosting device; S is the platform state at release as stored in the PCR inside the TPM; (Pu, Pr) is a non-migratable key pair such that the private part of the key Pr is bound to the TPM and to the platform state S. The following protocol functions are defined in [4]: $TPM_{CreateWrapKey}$, $TPM_{LoadKey2}$, TPM_{Seal}, and TPM_{Unseal}.

Algorithm 8.1 Server Agent Initialization

---

1. $M \rightarrow$ TPM: $TPM_{CreateWrapKey}$.
2. TPM: Generates a non-migratable key pair (Pu, Pr). Pr is bound to the TPM, and to the required platform state S at release, as stored in the PCR inside the TPM.
3. TPM $\rightarrow M$: TPM_KEY12[Pu, Encrypted Pr, TPM_KEY_STORAGE, tpmProof=TPM (NON-MIGRATABLE), S, Auth_data].

Algorithm 8.2 Administrators Registration

---

1. $M \rightarrow$ Administrators: Request for system administrators' authentication credentials.
2. $M \rightarrow$ TPM: $TPM_{LoadKey2}$(Pr). Loads the private key Pr in the TPM trusted environment, after verifying the current PCR value matches the one associated with Pr (i.e., S). If the PCR value does not match S, M returns an appropriate error message.
3. $M \rightarrow$ TPM: TPM_{Seal}(Authentication_Credential).

Algorithm 8.3 Authentication Verification

---

1. $M \rightarrow$ Administrators: Request for authentication credentials.
2. $M \rightarrow$ TPM: $TPM_{LoadKey2}$(Pr). TPM on M loads the private key Pr in the TPM trusted environment, after verifying the current PCR value matches the one associated with Pr (i.e., S). If the PCR value does not match S, M returns an appropriate error message.
3. $M \rightarrow$ TPM: TPM_{Unseal}(Authentication_Credential).
4. TPM: Decrypts the string Authentication_Credential and passes the result to M.
5. M: Authenticates the administrators using the recovered authentication credentials. If authentication fails, M returns an appropriate error message.

8.6.4   Client Agent Initialization

This section describes the procedure of initializing the organization and Cloud client agents. The goal of this procedure is to prepare devices to join a domain, which includes installing the client agent at a device. This covers generating the non-migratable key to protect domain credentials.

The protocol of initializing the client agent is described in Algorithm 8.4. The objective of this algorithm is to install at each device a copy of the client agent, which generates a non-migratable key to protect the domain credentials. Organization client agents are installed at member devices of the HD and OD, while Cloud client agents are installed at member devices of the MD (member devices of the COD and CMD are members of the OD and MD, so we do not need to install the client agents on them). The following notation is used in the provided algorithm: D is the client agent running on a client device; TPM, S, and (Pu, Pr) have the same meanings as provided earlier.

Algorithm 8.4   Client Agent Initialization

---

1. $D \rightarrow \text{TPM}_D$: $\text{TPM}_{CreateWrapKey}$.
2. TPM: Generates a non-migratable key pair (Pu, Pr).
3. TPM $\rightarrow D$: TPM_KEY12[Pu, Encrypted Pr, TPM_KEY_STORAGE, tpmProof=TPM (NON-MIGRATABLE), S, Auth_data].

8.7   Framework Workflow

This section discusses a possible workflow of the presented system framework.

8.7.1   Management Domain and Collaborating Management Domain Establishment

This procedure is followed when establishing an MD and CMD managed by the Cloud server agent. In the protocol we use the same notation described earlier (note that M resembles the Cloud server software agent running on the VCC).

In this subsection we require that the Cloud server agent has already been installed on the VCC, exactly as described earlier in Section 8.6.3. This includes installing the server agent, which interacts with the TPM to generate a non-migratable key pair that can only be used by the server agent. This key pair is used to protect the MD and CMD credentials.

Domain establishment begins when the Cloud administrators want to add a new domain to the Cloud infrastructure. Administrators instruct the server agent to create a new MD. The server agent authenticates the administrators, as described in Algorithm 8.3. If authentication succeeds, the server agent interacts with the TPM to securely generate an MD-specific secret key k_{md} and an identifier i_{md}, as described in Algorithm 8.5.

On successful completion of this protocol, the MD credentials are initialized at the VCC, including the MD key, MD identifier, and an empty PKL. These are protected by the VCC, which manages MD membership. Cloud internal employees (i.e., system architects) assign devices to the MD based on the required overall MD properties. They also define member device capabilities and policies governing their interaction. In addition, they join related MDs to a CMD. The policies control the migration of OD resources within a single MD resource and across multiple MD members of a CMD.

Algorithm 8.5 Management Domain Establishment

---

1. $M \rightarrow$ TPM: $\mathrm{TPM}_{GetRandom}$. TPM generates a random number to be used as MD domain key k_{md}.
2. TPM$\rightarrow M$: k_{md}.
3. $M \rightarrow$ TPM: $\mathrm{TPM}_{GetRandom}$. M generates a unique number to be used as MD domain identifier i_{md}.
4. TPM$\rightarrow M$: i_{md}.
5. The MD domain credentials k_{md}, i_{md}, and an empty PKL_{md} are stored in VCC protected storage and sealed to the server agent so that only the server agent can access these credentials when its execution status is trusted. This is achieved as follows:

   5a. $M \rightarrow$ TPM: $\mathrm{TPM}_{LoadKey2}(\mathrm{Pr})$. Loads the private key Pr in the TPM trusted environment to be used in the sealing function, after verifying the current PCR value matches the one associated with Pr (i.e., S). If the PCR value does not match S, M returns an appropriate error message.

   5b. $M \rightarrow$ TPM: $\mathrm{TPM}_{Seal}(k_{md}||i_{md}||\mathrm{PKL}_{md})$. TPM securely stores the string $k_{md}||i_{md}||\mathrm{PKL}_{md}$ using the platform protected storage, such that it can only be decrypted on the current platform by M, and only if the platform runs as expected (when the platform PCR values match those associated with Pr, i.e., S).

8.7.2 Organization Home Domain Establishment

This procedure is followed when an organization wishes its internal devices to access the outsourced applications at the Cloud. The process starts by creating an HD, which provides the following: enables controlled content sharing between member devices of the HD and other member devices of the OD, and simultaneously prevents uncontrolled transfer of content to other devices. The HD consists of all devices that are required to access or be accessible by the OD members. The HD is managed by the server agent running on the organization-specific master controller. The server agent initialization procedure is the same as that described in Section 8.7.1.

8.7.3 Adding Devices to a Domain

This section describes the process for adding a device to a domain (the process applies to all types of domain described in this chapter; note that collaborating domains (CMD and COD) are not included as they are covered by policies governing their members). The following notation is used in the provided protocol:

- D is the client agent running on a device.
- M is the server agent running on its hosting device (i.e., VCC for MD, MC for HD and OD).
- TPM_D is the TPM on a client device.

- TPM_M is the TPM on the server agent hosting device.
- S_D is the platform state at release as stored in the PCR inside the TPM_D.
- S_M is the platform state at release as stored in the PCR inside the TPM_M.
- $(\text{Pu}_D, \text{Pr}_D)$ are non-migratable key pairs such that the private part of the key Pr_D is bound to TPM_D and to the platform state S_D.
- $(\text{Pu}_M, \text{Pr}_M)$ are non-migratable key pairs such that the private part of the key Pr_M is bound to TPM_M and to the platform state S_M.
- i is the domain-specific identifier, which is equal to i_{md} for MD, i_{HD} for HD, or i_{OD} for OD.
- PKL is the domain public key list, which is equal to PKL_{md} for MD, PKL_{HD} for HD, or PKL_{OD} for OD.
- k is the domain-specific content protection key, which is equal to k_{md} for MD, k_{HD} for HD, or k_{OD} for OD.
- Cert_M is the server agent hosting device certificate.
- Cert_D is the joining device certificate.
- A_M is an identifier for the server agent device included in Cert_M.
- A_D is an identifier for a client agent device included in Cert_D.
- Pr_{MAIK} is the corresponding private key of the public key included in Cert_M.
- Pr_{DAIK} is the corresponding private key of the public key included in Cert_D.
- N_1 is a randomly generated nonce.
- N_2 is a randomly generated nonce.
- $e_{\text{Pu}_D}(Y)$ denotes the asymmetric encryption of data Y using key Pu_D, and where we assume that the encryption primitive in use provides non-malleability, as described in [7].

The client agent on the device sends a join domain request to the server agent to install the domain-specific key k. This request includes the domain-specific identifier i. Two algorithms are then initiated to add the device to the domain. The first algorithm involves the server agent and the client agent to mutually authenticate each other, conforming to the three-pass mutual authentication protocol [8]. The server agent sends an attestation request to the client agent to prove its trustworthiness, then the client agent sends the attestation outcome to the server agent. These steps are achieved using Algorithm 8.6.

Adding a device to a domain uses a second algorithm (Algorithm 8.7), which starts upon successful completion of Algorithm 8.6. The objective of Algorithm 8.7 is to securely transfer the key k to client agent D. k is sealed on D, so that it is only released to the client agent when its execution environment is as expected. If D's status is trusted, the server agent checks if the device's public key is included in the domain public key list. If so, it securely releases the domain-specific key k to D using Algorithm 8.7. The keys are sealed on D, so that they are only released to the client agent when its execution environment is as expected.

Upon successful completion of these algorithms, the client agent and the server agent establish a trusted secure communication channel that is used to transfer the domain key to the client agent. Such a channel provides assurance to the server agent about the client agent state, and also forces future use of the transferred key to the agent on specific trusted state. The device hosting D is now part of the domain, as it possesses a copy of the key k, and its public key matches the one stored in the server agent. Member devices of the domain can access the domain associated content, and hence such content is now shared by all member devices of the domain.

Algorithm 8.6   Client and Server Agents Mutual Authentication

---

1. $M \rightarrow \text{TPM}_M$: $\text{TPM}_{GetRandom}$.
2. $\text{TPM}_M \rightarrow M$: Generates a random number to be used as a nonce N_1.
3. $M \rightarrow \text{TPM}_M$: $\text{TPM}_{LoadKey2}(\text{Pr}_{MAIK})$; Loads the server agent hosting device AIK in the TPM trusted environment, after verifying the current PCR value matches the one associated with Pr_{MAIK}.
4. $M \rightarrow \text{TPM}_M$: $\text{TPM}_{Sign}(N_1)$.
5. $\text{TPM}_M \rightarrow M \rightarrow D$: $N_1 || \text{Cert}_M || \text{Sign}_M(N_1)$.
6. D: Verifies Cert_M, extracts the signature verification key of M from Cert_M, and checks that it has not been revoked, e.g. by querying an OCSP service [9]. D then verifies the message signature. If the verifications fail, D returns an appropriate error message.
7. $D \rightarrow \text{TPM}_D$: $\text{TPM}_{GetRandom}$.
8. $\text{TPM}_D \rightarrow D$: Generates a random number N_2 that is used as a nonce.
9. $D \rightarrow \text{TPM}_D$: $\text{TPM}_{LoadKey2}(\text{Pr}_{DAIK})$; Loads the private key Pr_{DAIK} in the TPM trusted environment, after verifying the current PCR value matches the one associated with Pr_{DAIK}.
10. $D \rightarrow \text{TPM}_D$: $\text{TPM}_{CertifyKey}(\text{SHA1}(N_2 || N_1 || A_M || i), \text{Pu}_D)$. TPM_D attests to its execution status by generating a certificate for the key Pu_D.
11. $\text{TPM}_D \rightarrow D$: $N_2 || N_1 || A_M || \text{Pu}_D || S_D || i || \text{Sign}_D(N_2 || N_1 || A_M || i || \text{Pu}_D || S_D)$.
12. $D \rightarrow M$: $N_2 || N_1 || A_M || \text{Pu}_D || S_D || i || \text{Cert}_D || \text{Sign}_D(N_2 || N_1 || A_M || i || \text{Pu}_D || S_D)$.
13. M verifies Cert_D, extracts the signature verification key of D from the certificate, and checks that it has not been revoked, e.g. by querying an OCSP service. M then verifies the message signature, the message freshness by verifying the value of N_1, and then verifies it is the intended recipient by checking the value of A_M. M determines if D is executing as expected by comparing the platform state given in S_D with the predicted platform integrity metric. If these validations fail, then M returns an appropriate error message.

Algorithm 8.7   Sealing Domain Key to Client Agent

---

1. $M \rightarrow \text{TPM}_M$: $\text{TPM}_{LoadKey2}(\text{Pr}_M)$. TPM on M loads the private key Pr_M in the TPM trusted environment after verifying the current PCR value matches the one associated with Pr_M (i.e., S_M). If the PCR value does not match S_M, the server agent returns an appropriate error message.
2. $M \rightarrow \text{TPM}_M$: $\text{TPM}_{Unseal}(k || i || \text{PKL})$.
3. $\text{TPM}_M \rightarrow M$: Decrypts the string $k || i || \text{PKL}$ and passes the result to M.
4. M verifies i matches the recovered domain identifier and Pu_D is included in the PKL. If so, M encrypts k using the key Pu_D as follows: $e_{\text{Pu}_D}(k)$.
5. $M \rightarrow \text{TPM}_M$: $\text{TPM}_{CertifyKey}(\text{SHA1}(N_2 || A_D || e_{\text{Pu}_D}(k)), \text{Pu}_M)$.
6. $\text{TPM}_M \rightarrow M$: Attests to its execution status by generating a certificate for the key Pu_M, and sends the result to M.
7. $M \rightarrow D$: $N_2 || A_D || \text{Pu}_M || S_M || e_{\text{Pu}_D}(k) || \text{Sign}_M(N_2 || A_D || e_{\text{Pu}_D}(k)) || \text{Pu}_M || S_M)$.
8. The device D verifies the message signature, that it is the intended recipient by checking the value of A_D, and verifies the message freshness by checking the value of N_1. If verifications succeed, D stores the string $e_{\text{Pu}_D}(k))$ in its storage.

8.7.4   Outsourced Domain and Collaborating Outsourced Domain Establishment

Section 8.1 discusses the general steps that an organization would follow when outsourcing part of (or all of) its infrastructure to the Cloud. This section assumes that an organization has chosen the applications to be outsourced to the Cloud, defined the requirements, and negotiated an SLA with the Cloud provider. The following is the process of establishing an OD and joining related ODs within a COD based on user requirements.

The organization would first need to communicate with the APIs of the Cloud provider, which create virtual resources considering the defined user requirements. Both endpoints (that is the Cloud provider and the organization) would need to validate each other's trustworthiness. The work on dynamic domains [10] has provided a protocol for establishing trusted secure channels between the endpoints of collaborating organizations. Such a trusted secure channel would provide the required assurance to such endpoints about each other's execution environment being trusted to behave as expected. The trusted secure channel establishes a trustworthy communication between the organization server agent running on the MC and the Cloud server agent running on the VCC.

The Cloud server agent and client agent have also established trusted secure channels, as discussed in Section 8.7.3. The secure channels between the organization and the VCC and between the VCC's server agent and the physical resources' client agents provide assurance to the organization that the channel from MC physical devices is trusted. The following steps are then executed:

Step 1. The organization server agent sends a request to the Cloud server agent to establish IaaS virtual resources. The request includes the organization technical requirements. Such requirements should be specified using a standard language, which is outside our scope to discuss.

Step 2. The Cloud server agent validates user properties. If validation succeeds, the Cloud server agent selects the CMD which could serve user properties. The selection of CMD would mainly be based on the infrastructure properties of MD members of the CMD.

Step 3. Based on user properties and the identified CMD, the Cloud server agent establishes user-specific processes and policies. These define how the CMD would manage the OD, for example, it defines for each OD the primary MD and backup MDs which should be members of the same CMD. Such processes and policies enable the primary MD to instantiate and control the user OD virtual resources. It also contains management decisions related to other OD members in the same COD.

Step 4. The Cloud server agent coordinates with the Cloud client agents of each identified MD to establish the MD allocated VMs.

Step 5. The Cloud client agents coordinate amongst themselves and create VMs, as defined by the user requirements.

Step 6. The Cloud server agent sends details of the newly created VMs (PKL, IP addresses, and default authentication credentials) to the organization server agent.

Step 7. The organization server agent creates a new OD following the same process as described in Section 8.7.2.

Step 8. The organization server agent then adds the created VMs to the OD exactly as described in Section 8.7.3. Following this process, member devices of the organization HD can communicate with the OD devices following the organization defined policy.

8.8   Discussion and Analysis

In this section we discuss the advantages of the framework architecture and how it achieves the objectives identified in Section 8.3.

8.8.1   Benefits of Using Trusted Computing

We use the 'remote attestation' concept in trusted computing, which provides the ability to remotely attest to the execution environment of running software agents (i.e., server and client agents). It also binds the release of domain credentials to the attested trusted environment. This provides assurance that agents are behaving as expected 'offline.' Such assurance establishes a *chain of trust* between an organization and the Cloud provider as follows:

- An organization server agent trusts a Cloud server agent to enforce user properties.
- The Cloud server agent trusts the Cloud client agent to enforce both user properties and infrastructure properties.
- The organization server agent trusts the organization client agent running at member devices of HD and OD to enforce user properties.

The first two points assure organizations that their client agents can enforce user properties, without worrying about the vulnerabilities caused by the Cloud – at this stage we do not consider attacks at the physical or the hypervisor level, as indicated in Section 8.3 and Assumption 8.1.

8.8.2   Benefits of the Framework Architecture

The framework architecture adds the following benefits:

- It delegates the management of OD to the owner organization rather than to the Cloud provider; in addition to the established chain of trust described above, this enables organizations to have control over their outsourced applications at the Cloud.
- The Cloud provider delegates the enforcement of a domain policy to the organization. Similarly, the Cloud server agent delegates the enforcement of the domain policy to the Cloud client agents. If all resources must be fully managed at all times by a centralized management unit, then this could be subject to single point of failure and would also raise performance concerns.
- The framework supports domains with special properties (e.g., expandability, changeability of member devices, and collaboration with other domains). As the Cloud environment is dynamic, this feature will be extremely helpful in satisfying Cloud properties and management requirements defined earlier, especially for OD and MD.
- Secure domains enable controlled content sharing but protection between member devices of a domain.

8.8.3   Content Protection

Protecting organization data and Cloud management data from Cloud insiders is achieved as follows (as discussed earlier, we do not focus on hypervisor security threats and physical security threats in this chapter). If an authorized or unauthorized insider sends content from a member device of an organization OD to an unauthorized user, the content stays protected and the unauthorized user will not be able to access content on any device which is not a member of the OD. This is because client agents are trusted to not reveal protected content to others. Thus, if an insider transfers protected content to another device which is not a member of the OD, the receiver will not be able to access the protected content as the receiver does not possess a copy of the content decryption key.

If an insider attempts to send a copy of the OD key to unauthorized users they will fail to do so. This is because the organization server and client agents are the only entities authorized to access the key after verifying the execution environment state of the device matches the one associated with the keys. In other words, these keys are sealed to be used only by a trusted application, which is implemented to not reveal the keys in the clear even to system administrators.

If an insider attempts to add an unauthorized device to the OD they will fail to do so. This is because system administrators explicitly identify member devices of a domain by adding their public keys in a domain-specific public key list. This means only predefined devices can join a domain. Therefore, unauthorized devices will fail to join a domain as their public key is not listed in the PKL and so they will not be able to get a copy of the domain key.

From the above we can see that the discussed scheme enables controlled content sharing between devices. Simultaneously, content can only be accessed by devices which are authorized by security administrators. The same discussion applies to all other types of domain but to address different security threats. For example, an insider might add an insecure device to the MD, migrate a VM to the new device, and then leak content. Our scheme addresses such a threat by controlling the member devices of an MD as discussed above.

8.9 Summary

Trust establishment in Clouds requires collaborative efforts from industry and academia. Establishing trust in Cloud systems requires two mutually dependent elements: (a) supporting infrastructures with trustworthy mechanisms and tools to help Cloud providers automate the process of managing, maintaining, and securing their systems (this includes but is not limited to self-managed services); and (b) developing methods to help Cloud users and providers to establish trust in the operation of the infrastructure by continually assessing its operational status. This chapter focuses on point (b) and, in addition, it discusses the way to transfer the power to manage users' application data from the hand of the Cloud providers to the hand of the users.

The framework presented in this chapter presents a foundation roadmap, but it is not enough by itself to establish an end-to-end trusted Cloud. That is, the framework still requires further extensions as establishing trust in Clouds is a complex subject. Chapter 10 extends the framework to cover Cloud provenance.

8.10 Exercises

Q1. Discuss the main challenges that would need to be considered when an organization, which has a private Cloud, outsources its services to a public Cloud model.

Q2. Discuss the main challenges that would need to be considered when an organization, which has a private Cloud, outsources part of its services to a community Cloud model.

Q3. Discuss the main challenges that would need to be considered when an organization, which has a private Cloud, outsources part of its services to a hybrid Cloud model.

Q4. Discuss what organizations could do to establish trust in Clouds in each of the above cases, i.e. Q1, Q2, and Q3.

References

[1] Imad M. Abbadi. Operational trust in clouds' environment. In *MoCS 2011: Proceedings of the Workshop on Management of Cloud Systems*, pp. 141–145. IEEE, June 2011.

[2] Richard Chow, Philippe Golle, Markus Jakobsson, Elaine Shi, Jessica Staddon, Ryusuke Masuoka, and Jesus Molina. Controlling data in the cloud: Outsourcing computation without outsourcing control. In *Proceedings of the 2009 ACM Workshop on Cloud Computing Security, CCSW '09*, pp. 85–90. ACM: New York, 2009.

[3] Keith Jeffery and Burkhard Neidecker-Lutz. The Future of Cloud Computing – Opportunities for European Cloud Computing Beyond 2010.

[4] Trusted Computing Group. *TPM Main, Part 2, TPM Structures. Specification version 1.2 Revision 103*, 2007.

[5] Muntaha Alawneh and Imad M. Abbadi. Sharing but protecting content against internal leakage for organisations. In *DAS 2008*, vol. 5094 of *LNCS*, pp. 238–253. Springer-Verlag: Berlin, 2008.

[6] Jonathan M. McCune, Yanlin Li, Ning Qu, Zongwei Zhou, Anupam Datta, Virgil D. Gligor, and Adrian Perrig. Trustvisor: Efficient TCB reduction and attestation. In *IEEE Symposium on Security and Privacy*, pp. 143–158, 2010.

[7] International Organization for Standardization. *ISO/IEC 18033-2, Information technology – Security techniques – Encryption algorithms – Part 2: Asymmetric ciphers*, 2006.

[8] International Organization for Standardization. *ISO/IEC 9798-3, Information technology – Security techniques – Entity authentication – Part 3: Mechanisms using digital signature techniques*, 2nd edn, 1998.

[9] M. Myers, R. Ankney, A. Malpani, S. Galperin, and C. Adams. X.509 Internet Public Key Infrastructure Online Certificate Status Protocol – OCSP. RFC 2560, Internet Engineering Task Force, June 1999.

[10] Muntaha Alawneh and Imad M. Abbadi. Preventing information leakage between collaborating organisations. In *ICEC '08: Proceedings of the Tenth International Conference on Electronic Commerce*, pp. 185–194. ACM: New York, 2008.

9

Clouds Chains of Trust

This chapter establishes a foundation framework which draws a roadmap for addressing the first two challenges discussed in Chapter 7; that is, developing the effective chain of trust functions and the dynamicity aware protocols. The framework addresses the question of how an entity could establish trust in a composition of multiple entities which could change dynamically. The chapter also discusses how users could assess the Cloud's trustworthiness without the need to get involved in the complex technical details of the Cloud.

9.1   Introduction

Establishing trust between remote entities is an important subject which has been widely discussed in academia and industry. The most commonly known solution attempting to address this problem, which has been adopted by the industry, is that proposed as part of the TCG specifications (known as remote attestation) [1–3]. Establishing remote attestation in Clouds is critical for their success. However, remote attestation as proposed by the TCG is impractical in the Cloud due to Clouds' complexity and dynamism. This chapter clarifies this important subject, and presents a method which helps in providing remote attestation in the context of the Clouds environment.

Establishing trust in the Cloud infrastructure is an important subject that is yet to receive adequate attention from both academia and industry [4–7]. There are a number of techniques that enable one party to establish trust in an unknown entity: direct interaction, trust negotiation, reputation, and trust recommendation and propagation. Most of these establish trust based on identity. Trust negotiation, by contrast, establishes trust based on properties. In a Cloud context, establishing trust would be based on both identities and properties [8]. This chapter focuses on *trust negotiation* based on properties. Specifically, it presents enhancements to remote attestation [1–3] which provide a *trustor* with an authentic and fresh copy of the properties of the *trustee* such that the *trustor* can make a timely decision on the ability of the *trustee* to operate in a certain state.

Remote attestation, as discussed by the TCG specifications, requires the *trustee* to have a root of trust [1–3]. The root of trust must be trusted by the *trustor* (e.g., a third party vouches

Cloud Management and Security, First Edition. Imad M. Abbadi.
© 2014 John Wiley & Sons, Ltd. Published 2014 by John Wiley & Sons, Ltd.
Companion Website: www.wiley.com/go/abbadi_cloud

for its trustworthiness). In TCG, the root of trust is protected by a tamper-evident hardware chip (the TPM) from where the root of trust starts. Remote attestation was originally designed to work between peers of devices. However, Clouds have different properties than peer-to-peer communication, as the building up of its resources – starting from a physical resource, hosting a virtual resource, which in turn runs an application resource – is dynamic. Chapter 7 demonstrates, using several examples, the effects of the Cloud dynamic nature on 'breaking' the trust relationship between a *trustee* and a *trustor* – any established relationship could be invalidated at any time as a result of the Cloud dynamic nature. The focus of this chapter is on the two identified agendas in Chapter 7: the compositional chains of trust and the transparency versus trust evaluation.

The chapter is organized as follows. Section 9.2 briefly summarizes the software agents which are discussed in Chapter 8. Section 9.3 defines the chain of trust concept and discusses its types. Section 9.4 presents the Clouds compositional chains of trust within a layer. Section 9.5 discusses the chains of trust across layers and how the chains of trust help in establishing Cloud 'trust anchors.' Finally, Section 9.6 summarizes the chapter.

9.2   Software Agents Revision

As we discussed earlier, trust establishment in Clouds requires 'trustworthy' automated self-managed services that can manage the Cloud infrastructure with minimum human intervention. Chapter 8 presents a trust establishment framework which identifies the challenges and requirements, and addresses those related to establishing a secure environment at the infrastructural level. This includes the establishment of offline chains of trust amongst the entities of the Cloud. This chapter builds on the discussed framework and addresses the question of how a verifier could attest to the trustworthiness of the complex and dynamic Cloud environment. The functions of the previously presented framework are provided using two types of software agent:

- A server software agent that runs at the VCC. We refer to the server software agent as the domain controller server side, DC-S.
- A client agent that runs at the physical resources within the Cloud infrastructure. We refer to the client software agent as the domain controller client side, DC-C.

The DC-S delegates the enforcement of some Cloud policies to DC-Cs. Each DC-C is in charge of enforcing the delegated policy. Prior to policy delegation, the DC-S establishes chains of trust with each of the DC-Cs, as discussed in the previous chapter. The following is an outline summary:

- A DC-S verifies the DC-C trustworthiness to continually enforce the domain policies and to only access the domain credentials when the resource execution status is as expected.
- In turn, the DC-C provides assurance to the DC-S about the trustworthiness of its hosting resource's execution environment when managing the domain and enforces the domain policies. This provides the assurance that only resources with a trustworthy DC-C can be members of a domain.

9.3   Roots of and Chains of Trust Definition

This section provides basic background about the roots and chains of trust.

9.3.1   Roots of Trust

In this part we first briefly highlight the Cloud taxonomy, which was covered in detail in the first part of this book, and then indicate the roots of trust within the taxonomy. The Cloud environment is composed of enormous *resources*, which are categorized based on their types and deployment across the Cloud infrastructure. A *resource* is a conceptual entity that provides services to other entities. The Cloud environment conceptually consists of multiple intersecting layers, as follows:

- *Physical layer*. This represents the physical resources and their interactions, which constitute the Cloud physical infrastructure. Examples of these resources include server, storage, and network resources. The physical layer resources are consolidated to serve the *virtual layer*.
- *Virtual layer*. This represents the virtual resources, which are hosted by the *physical layer*. Examples of these resources include VM, virtual network, and virtual storage. Cloud customers in an IaaS Cloud type interact directly with the virtual layer resources, which host Cloud customer applications.
- *Application layer*. This runs the applications of the Cloud customers. These are hosted using *virtual layer* resources. Cloud customers using a PaaS type deploy their applications at virtual layer resources, while customers of Cloud SaaS type access a deployed application via the Internet.

Figure 9.1 provides a conceptual model in which we identify an entity layer as the parent of the three Cloud layers (i.e., physical, virtual, and application). At an abstract level the layers contain resources which join domains (i.e., we have a physical domain, a virtual domain, and an application domain). A domain resembles a container which consists of related resources.

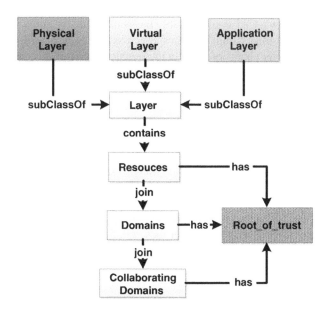

Figure 9.1   Cloud computing – layering conceptual model

Domain resources are managed following the domain defined policy. Domains that need to interact amongst themselves within a layer join a collaborating domain (i.e., we have a physical collaborating domain, a virtual collaborating domain, and an application collaborating domain). A collaborating domain controls the interaction between domains which are members of the collaborating domain using a defined policy.

The nature of resources, domains, collaborating domains, and their policies is layer specific. The concepts of domains and collaborating domains help in managing the Cloud infrastructure, and managing resources distribution and coordination in normal operation and during incidents. Collaborating domains communicate across Cloud layers to serve the collaborative customer application needs. Domains communicate horizontally within a layer-specific collaborative domain, and/or vertically across multiple layers of collaborative domains.

Each of the identified Cloud entities has a *root of trust* which helps in establishing trust in Clouds. Subsequent sections clarify the relationships between Cloud entities, define the roots of trust, and discuss how they would help in establishing trust in Clouds.

9.3.2   Chains of Trust

A chain of trust (CoT) is composed of a set of elements primarily used to establish the trust status of an object. The first element of the CoT (also call the *root of trust*) should be established from a trusted entity or an entity that is assumed to be trusted, for example a trusted third party or a tamper-evident hardware chip (as in the case of TPM) [1–3]. The trust status of the second element in the CoT is measured by the root of trust (i.e., the first element in the CoT). If the verifier trusts the root of trust, then the verifier must also trust the root of trust measurement of the second element. The second element then measures the trust status of the third element in the CoT. If the second element is trusted, and the second element measures the third element trust status, then the verifier trusts the measurements of the third element. This process is a simplified example of how a CoT could possibly be established.

Clouds have chains of trust within a layer and across layers. The across-layers chains of trust build on the intra-layer chains of trust. In the remaining part of this chapter we focus on these. The intra-layer chain of trust has two types:

- A single resource CoT.
- A compositional CoT representing multiple entities (i.e., domains and collaborating domains).

As we discussed in Chapter 7, a verifier is mainly interested in evaluating compositional CoTs within and/or across Cloud layers without the need to get involved in understanding the details of the Cloud infrastructure. The compositional CoT would be built on resource CoTs. As a result, this chapter defines both types of CoT, including defining the nature of their roots of trust (Figure 9.3 later illustrates the details of these relations). Subsequently, we move on to discuss Chains of Trust across the Cloud layers.

9.4   Intra-layer Chains of Trust

In this section we discuss a single resource CoT and a compositional CoT within a Cloud layer.

9.4.1   A Resource Chain of Trust

As stated earlier, a resource is a conceptual entity that provides services to other entities. Therefore, we begin the discussion by defining the CoT for a single resource (RCoT) as a triple comprising an initial trust function (itf), a set of trust functions (stf), and a sequence of elements in the chain ($sq. < x_0, x_1, \ldots, x_n >$), where x is an element representing any component (software, hardware, etc.) that contributes to the chain of trust. RCoT requires the following:

- The initial function evaluates to 'trusted' or 'assumed to be trusted' when applied to the first element of the sequence.
- Every function in the set of trust functions evaluates to 'true' when applied to any two consecutive elements of the sequence.

This is formally defined as follows:

$$RCoT = ($$
$$itf, stf, sq. < x_0, x_1, \ldots, x_n > \mid$$
$$itf(x_0) \in \{trusted, assumed\_trusted\},$$
$$\forall i : [1..n] \bullet \forall f : stf \bullet f(x_{i-1}, x_i) == true)$$

The nature of the $root\_of\_trust$ (i.e., the first element in the sequence, x_0) is based on the type of entity and its location within the Cloud layers. We now discuss a single resource $root\_of\_trust$ and the subsequent part of this section covers a compositional entities $root\_of\_trust$. We now clarify RCoT in the context of TCG specifications, as we require each resource within the physical layer to be TCG compliant and fitted with a TPM which is physically bound to that resource. A TPM must be tamper-evident; that is it provide a limited degree of protection against physical attack. The TPM helps in providing three roots of trust, for measurement, storage, and reporting. The RTM is a computing engine capable of making reliable measurements of TP running components, which is known as an integrity measurement. Integrity measurement is a cryptographic digest or hash of a TP component; that is, a piece of software executing on a TP [9]. The RTS is a collection of capabilities, which must be trusted if the storage of data in a TP is to be trusted [10]. The RTS uses TPM components to achieve its functions. The RTR is a collection of capabilities that must be trusted if reports of integrity metrics are to be trusted (platform attestation) [10]. The RTR works in conjunction with the RTM and the RTS to implement the platform attestation. The RTR enables a TPM to reliably report information about its identity and the current state of the TPM host platform. This is achieved using a set of keys and certificates, which are signed by a variety of third parties that must be trusted if the state of the platform is to be trusted. In TCG, the RCoT starts from the (CRTM, which should be stored in protected location such as the TPM (currently it is protected by the BIOS). Once the CRTM measures the initial environment state it stores the result in protected registers inside the TPM (referred to as PCR). The CRTM represents the $root\_of\_trust$, x_0, and the Set.(trust_functions) contains RTM, RTS, RTR, and other functions. The $initial\_trust\_function$ is the one that measures the CRTM itself and stores the result inside the TPM PCR. Figure 9.2 illustrates these relations.

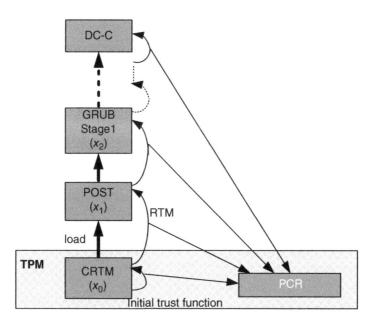

Figure 9.2   Example of a partial RCoT of a physical resource

Unlike the RCoT at the physical layer, RCoTs at the virtual and application layers have different treatments when discussing specific resource roots of trust. This is because physical resources are the foundation of virtual resource roots of trust, which in turn form the foundation of application resource roots of trust. In other words, the virtual and application layer RCoTs, in a Clouds context and considering its dynamisms, should build on a compositional CoT and not a specific RCoT.

We require each resource within the virtual layer to have a vTPM (virtual TPM). The vTPM provides similar roots of trust and trust functions as those described in the TPM. The main difference is that the vTPM is a software component and would need to assure its own trustworthiness. The trustworthiness of the vTPM used to be assured by extending the CoT from the hardware TPM up to the vTPM. However, in a Clouds context, this does not work because of the dynamic nature of Clouds. Therefore, we present in this chapter a different method in which the vTPM extends its trustworthiness from a compositional CoT at the physical layer rather than a specific RCoT at the physical layer. In other words, the virtual RCoT will start from the vTPM *root_of_trust* supported by the serving physical layer compositional CoT as discussed later in this chapter. Similar to a virtual RCoT, an application RCoT *root_of_trust* would typically start from the vTPM. However, in a Clouds context, an application resource is likely to be hosted by multiple virtual resources which increase/decrease based on demand (i.e., as a result of the elasticity property of Clouds). Therefore, having the *root_of_trust* of an application resource start from the vTPM would create complexities in calculating the effective chain of trust viewed by a user, as discussed in [11]. As a result, an application RCoT will represent the elements of the application resource supported by a *root_of_trust* resulting from the virtual layer compositional CoT, as discussed later in this chapter.

Having discussed different layer RCoTs, we now discuss the operation we use when discussing compositional CoTs.

- *compare($RCoT_1, RCoT_2$)*: Returns true if the CoTs are equal, otherwise returns false. This function is formally defined as follows:

$$compare(X, Y) = if\ RCoT_1 == RCoT_2\ then\ true$$
$$elsefalse$$

- *extend($RCoT_1, < elements >$)*: Concatenates $< elements >$ to $sq.RCoT_1$. Formally, this is defined as follows:

$$extend(X, Y) = (itf.X, stf.X, sq.X \frown Y)$$

- *combine($RCoT_1, RCoT_2$)*: Returns a unique sequence of elements, each representing the input CoT. Formally, this is defined as follows:
  $$combine(X, Y) = \{X, Y\}$$
  This function has the following semantics:
  - Idempotent: The combination of a chain of trust with itself or its equivalent is the chain of trust itself. Formally, this is defined as follows:
  $$combine(X, X) = \{X, X\} = \{X\}$$
  - Commutative: The resulting chain of trust from a combination does not depend on the order of combination:
  $$combine(X, Y) = combine(Y, X)$$
  - Associative: The grouping of combinations of chains of trust has no effect on the resulting chain of trust.
  $$combine(combine(X, Y), Z) = combine(X, combine(Y, Z))$$
  For simplicity, we write $combine(X, Y, Z)$ to mean either way.

9.4.2   Compositional Chains of Trust

We now define how trust is composed from members of a particular grouping in Clouds. Understanding compositional chains of trust is a vital requirement for establishing trust in Clouds. This is because Cloud resources at the upper layers are served by a collaborating set of resources rather than a specific resource. We identify two types of domain configuration:

- *Homogeneous domains.* In a homogeneous setting all resources are configured uniformly, resulting in identical CoTs. An example of this is the resources within a physical domain or a virtual domain. Each resource member of a physical domain is identical and carefully selected, interconnected, and positioned to achieve the domain properties. Similarly, resources of a virtual domain are identical as they represent (as a result of horizontal scalability) a replication of the VMs hosting an instance of an application resource.
- *Heterogeneous domains.* In a heterogeneous setting resources does not necessarily need to be configured uniformly, which results in differences in the CoTs. Application domains are an example of heterogeneous settings as they are composed of resources having different CoTs.

Collaborating domains follow the same concept as domains. For example, collaborating domains of the physical layer are homogeneous as they should serve as a backup for each other. Virtual and application layer collaborating domains, however, are heterogeneous as they serve to identify the interdependencies between domains rather than as a backup for each other. Chapter 2 explicitly identifies the trust relationship between Cloud entities.

We identify two types of compositional CoT, namely the domains chain of trust (DCoT) and the collaborating domains chain of trust (CDCoT). These CoTs are composed of two entities:

- A *root_of_trust*.
- A combination of all CoTs of the entities which are members of the domain and/or the collaborating domain.

Unlike a RCoT, the *root_of_trust* of DCoT/CDCoT attests to the trustworthiness of the way the domain or collaborating domain is managed and operated. We need a *root_of_trust* that satisfies two main properties: its trustworthiness can be measured and assessed at all times; and it can provide strong assurance about the trustworthiness of the way the domain or collaborating domain is managed and operated. In this section we discuss the compositional CoT in the following order: the physical layer, the virtual layer, and finally the application layer. We follow this order as each layer *root_of_trust* is derived from the layer immediately underneath it.

9.4.3   Physical Layer DCoT and CDCoT

We identify two types of compositional CoT at the physical layer, namely DCot and CDCoT. In these, a CoT is composed of two entities: a *root_of_trust* and a combination of the entities' CoTs which are members of the corresponding domain and/or the collaborating domain. Unlike RCoT, the *root_of_trust* of DCoT/CDCoT attests to the trustworthiness of the way the domain or collaborating domain is managed and operated. We need a *root_of_trust* that satisfies two properties:

- Its trustworthiness can be measured and assessed at all times.
- It can provide strong assurance about the trustworthiness of the way domains and collaborating domains are managed and operated.

Considering the above discussion, the two elements of the physical layer DCoT are as follows:

- *The combination of all RCoT members of a physical domain.* The physical domain is homogeneous and, as a result, all RCoT members of the domain are identical (combining identical chains of trust is equal to either chain of trust, as discussed in Section 9.4.1).
- *The root_of_trust of a physical domain.* As explained in Section 9.2, the resources of a physical domain run a trustworthy copy of DC-C which provides assurances of the physical resource state. The VCC runs a trustworthy copy of DC-S which measures the trustworthiness of DC-C. A verifier can independently acquire the CoT of the VCC and assess its trustworthiness; that is, a verifier can attest to the trustworthiness of DC-S which itself attests to the trustworthiness of DC-C.

These two points satisfy our two stated properties of DCoT/CDCoT *root_of_trust*. Therefore, we present the CoT of VCC, $RCoT(VCC)$, to act as the *root_of_trust* of the physical DCoT.

Assume a homogeneous physical domain, $D_{Physical}$, consists of resources $R_0, R_1, ..., R_n$ such that $\forall i, j : [0..n] \bullet RCoT(R_i) == RCoT(R_j)$. The $DCoT(D_{Physical})$ is then defined as follows:

$$DCoT(D_{Physical}) = combine(RCoT(VCC), RCoT(R_0))$$

DC-S, which is part of RCoT(VCC), vouches for and attests to the trustworthiness of members of $D_{Physical}$. DC-S also provides the assurance that DC-C can only operate and be a member of a domain when its serving host has a specific RCoT. Therefore, a verifier only needs to attest to the trustworthiness of RCoT(VCC) and DC-C, that is an extended CoT which starts from RCoT(VCC) and extends to DC-C. The resources of each physical domain have identical values of DC-C when it runs as expected.[1] As a result, we can redefine $DCoT(D_{Physical})$ as follows:

$$DCoT(D_{Physical}) = extend(RCoT(VCC), DC - C)$$

After discussing a physical layer DCoT we now move to the physical layer CDCoT. DC-S and DC-C manage both physical domains and physical collaborating domains. As a result, an appropriate *root_of_trust* of CDCoT is the same as the *root_of_trust* of DCoT. The *root_of_trust* of CDCoT is already included in DCoT, thus we can exclude it from the physical CDCoT. Suppose a collaborating domain $CD_{Physical}$ is composed of domains $D_{P0}, D_{P1}, ..., D_{Px}$ such that $\forall i, j : [0..x] \bullet DCoT(D_{Pi}) == DCoT(D_{Pj})$. The $CD_{Physical}$ is then defined as follows:

$$CDCoT(CD_{Physical}) = DCoT(D_{P0})^2$$

By substituting the value of $DCoT(D_{P0})$ from $DCoT(D_{Physical})$, we end up with the following:

$$CDCoT(CD_{Physical}) = extend(RCoT(VCC), DC - C)$$

The above shows that physical CDCoT is mainly based on VCC and DC-C. VCC trustworthiness can be measured by a verifier, and DC-C trustworthiness can be verified by VCC. This is the foundation of the *root_of_trust* of the physical layer, which acts as a foundation for the layer above it (i.e., the virtual layer), as discussed in [12].

9.4.4 Virtual Layer DCoT and CDCoT

Having defined the DCoT and CDCoT at the physical layer, we now move to the virtual layer. A virtual resource can only be served by resource members of a specific physical collaborating domain (as discussed in Chapter 2). Therefore, we present the *root_of_trust* of a virtual DCoT to be built on the hosting physical CDCoT, that is $extend(RCoT(VCC), DC - C)$. The *root_of_trust* still satisfies our stated properties, as DC-Cs running at the physical collaborating domain manage the virtual resources and measure their trustworthiness. DC-C trustworthiness

---

[1] Resources of different domains would typically have different values of DC-C, as different domains have different properties.
[2] As discussed earlier, each physical collaborating domain has identical physical domains to support Cloud properties; however, resources and domain members of different collaborating domains are not necessarily identical.

is measured and assured by DC-S. DC-S trustworthiness can be measured by a verifier. This builds a CoT from DC-S to DC-C, and from DC-C to the virtual resources. As in the case of the physical layer the Cloud elasticity property results in each virtual domain's resources having identical RCoTs. A virtual domain, $D_{Virtual}$, DCoT is then defined as follows. Assume that a physical collaborating domain, $CD_{Physical}$, whose CDCoT is $CDCoT(CD_{Physical})$, hosts a virtual domain $D_{Virtual}$. Assume $D_{Virtual}$ has resources $V_0, V_1,..., V_k$ such that $\forall a, b : [0..k] \bullet RCoT(V_a) == RCoT(V_b)$. The $DCoT(D_{Virtual})$ is then defined as follows:

$$DCoT(D_{Virtual}) = combine(extend(RCoT(VCC), DC - C), RCoT(V_0))$$

Unlike virtual domains, virtual collaborating domains consist of virtual domains that support different services. Such services interact amongst each other to serve the needs of an application domain. Domains that provide different services would typically have different DCoTs. As in the case of physical CDCoT, the *root_of_trust* of the virtual CDCoT is included in its member domains, and so we do not need to re-include it. Suppose a virtual collaborating domain $CD_{Virtual}$ is composed of domains D_{V0}, D_{V1}, D_{V2} such that $\forall i, j : [0..2] \bullet DCoT(D_{Vi})! = DCoT(D_{Vj})$. The following is the definition of a CDCoT of non-identical virtual domains:

$$CDCoT(CD_{Virtual}) = combine(DCoT(D_{V0}), DCoT(D_{V1}), DCoT(D_{V2}))$$

By substituting the values of $DCoT(D_{V0})$, $DCoT(D_{V1})$, and $DCoT(D_{V2})$, and assuming $Vx_0 \in D_{V0} \wedge Vy_0 \in D_{V1} \wedge Vz_0 \in D_{V2}$, we end up with the following:

$CDCoT(CD_{Virtual}) = combine($
$combine(extend(RCoT(VCC), DC - C), RCoT(Vx_0),$
$combine(extend(RCoT(VCC), DC - C), RCoT(Vy_0),$
$combine(extend(RCoT(VCC), DC - C), RCoT(Vz_0))$
$\equiv$
$CDCoT(CD_{Virtual}) = combine(extend(RCoT(VCC), DC - C),$
$RCoT(Vx_0), RCoT(Vy_0), RCoT(Vz_0))$

9.4.5 Application Layer DCoT and CDCoT

An application domain is composed of multiple resources where each resource provides part of the functions that other resources depend on. As a result, the application domain has a heterogeneous mix of RCoTs. The effective application DCoT results in a complicated structure, and its *root_of_trust* is built on a collaborating domain chain of trust of the hosting layer which is the virtual CDCoT. Application resources are of two types:

- Application resources which provide the same functions. In this case such applications would typically have the same components. This results in a symmetric RCoT between the components. Such symmetry reflects various real-life scenarios, as in the case of replicated web applications and replicated database management systems
- Application resources which provide different functions. An example of this are the resources of dependent applications which would typically have different components resulting in differences in their RCoT (e.g., resources providing web applications that connect and build on resources providing database management systems).

The dynamic nature of the Cloud, as discussed earlier, could result in users being connected to an application resource at one time but served by a replicated resource member of the same application domain at a different time. A verifier, therefore, is interested in assessing the trustworthiness of a whole domain.

An application RCoT would build on a specific virtual DCoT, as each application resource would be served by a specific virtual domain (see Chapter 2). All virtual domains serving a specific application domain would join a virtual collaborating domain. As a result, the application domain *root_of_trust* should be the virtual CDCoT.

Suppose an application domain, D_{App}, is composed of resources $Ax_0, Ax_1, \ldots, Ax_l, Ay_0, Ay_1, \ldots, Ay_m, Az_0, Az_1, \ldots, Az_n$ such that

$$\forall i,j : [0..l] \bullet RCoT(Ax_i) == RCoT(Ax_j) \wedge$$
$$\forall i,j : [0..m] \bullet RCoT(Ay_i) == RCoT(Ay_j) \wedge$$
$$\forall i,j : [0..n] \bullet RCoT(Az_i) == RCoT(Az_j) \wedge$$
$$RCoT(Ax_0) \,! = RCoT(Ay_0) \wedge RCoT(Ax_0) \,! = RCoT(Az_0) \wedge$$
$$RCot(Ay_0) \,! = RCot(Az_0)$$

The following is the definition of a $DCoT(D_{App})$:

$$DCoT(D_{App}) = combine(CDCoT_{Virtual}, RCoT(Ax_0), RCoT(Ay_0), RCoT(Az_0))$$

An application collaborating domain would typically consist of domains that provide different services. Such domains join a collaborating domain to have a policy governing their interdependencies, that is they should be located within physical proximity, have less restricted access between their resources, etc. As in the case of virtual domains, application domains would typically have different DCoTs. Suppose a collaborating domain, CD_{App}, is composed of domains D_{A0}, D_{A1}, and D_{A2}. The following is the definition of an application CDCoT (we exclude the *root_of_trust* as it is already included in the individual domain DCoTs):

$$CDCoT(CD_{App}) = combine(DCoT(D_{A0}), DCoT(D_{A1}), DCoT(D_{A2}))$$

We now use the definition of $DCoT(D_{App})$ to re-build $CDCoT(CD_{App})$. Assume the following: D_{A0} consists of two unique resources, Ax_0 and Ax_1, which are served by a virtual collaborating domain whose CDCoT is $CDCoT_{Virtualx}$; D_{A1} consists of two unique resources, Ay_0 and Ay_1, which are served by a virtual collaborating domain whose CDCoT is $CDCoT_{Virtualy}$; and D_{A2} consists of two unique resources, Az_0 and Az_1, which are served by a virtual collaborating domain whose CDCoT is $CDCoT_{Virtualz}$. The definition of $CDCoT(App)$ is then extended as follows:

$$CDCoT(App) =$$
$$combine(combine(CDCoT_{Virtualx}, RCoT(Ax_0), RCoT(Ax_1)),$$
$$combine(CDCoT_{Virtualy}, RCoT(Ay_0), RCoT(Ay_1)),$$
$$combine(CDCoT_{Virtualz}, RCoT(Az_0), RCoT(Az_1)))$$
$$=$$
$$combine(CDCoT_{Virtualx}, CDCoT_{Virtualy}, CDCoT_{Virtualz},$$
$$RCoT(Ax_0), RCoT(Ax_1), RCoT(Ay_0),$$
$$RCoT(Ay_1), RCoT(Az_0), RCoT(Az_1))$$

9.5   Trust Across Layers

In this section we discuss how the identified trust relations and chains of trust help in establishing the 'trust anchors' of the Cloud. Trust anchors aim to provide verifiers with mechanisms enabling them to transparently evaluate trust in the Cloud at different layers of abstractions. The level of transparency is directly proportional to the type of verifier. For example, if a verifier is an IaaS Cloud user then he would need to be provided with mechanisms assuring him of the trustworthiness of the Cloud to manage his virtual resources at physical resources (as agreed in the SLA). If a verifier is a SaaS user then he would need to be provided with mechanisms assuring him of the trustworthiness of the Cloud to manage software applications. This would cover the hosting environment (physical and virtual), as agreed in the SLA. If the verifier is a PaaS user then he would need two assurances: the same assurances as IaaS, and an additional assurance of the trustworthiness of the Cloud to manage all additional software components of the development environment provided by the Cloud, as required by the application of the verifier.

As discussed earlier, layers in a Cloud serve as a means of abstracting internal details of its complex infrastructure and resources. Each layer creates a pool of entities that can be used by the layer above it. Therefore, an application, user, or process at layer (n) is presented with a pool of entities by layer $(n - 1)$. Layer (n), as a result, assumes certain aspects about the trustworthiness of the pool provided by layer $(n - 1)$. Based on the level of trust it builds services on top of this pool of resources. We therefore state that trust in the Cloud at any layer depends on the trustworthiness of the pool of resources provided by the layer underneath it.

Providing transparent trust management involves two conflicting requirements:

- Provide a verifier with measurements of the pool of resources at layer (n) and simultaneously assure the verifier about the trustworthiness of the pool of resources at layer $(n - 1)$ (i.e., the one that hosts the verifier resources).
- The verifier should not get involved in understanding and measuring the complex trust relations amongst and across Cloud serving layers.

This chapter answers the question: *Can TCG specifications be used to address the stated requirements?* In this chapter we discuss the fact that the current TCG concepts, such as remote attestation, do not work 'as is' in Clouds because of their dynamic nature. In addition, remote attestation requires the *trustee* to reveal the whole chain of trust of all physical resources at different layers of abstraction. As a result, the *trustor* would need to understand the details of the Cloud infrastructure. Therefore, using TCG concepts 'as is' does not satisfy the stated objectives.

The method discussed throughout this chapter still builds on TCG concepts; however, it moves the starting point of the root of trust from the TPM to the discussed trust anchors. The trust anchors are layer specific and they generally represent a group of compositional resources at each layer. Different types of Cloud users have different interest in the Cloud layers – see the three solid arrows in Figure 9.3, which clarify the differences in user requirements for assessing resources at Cloud layers. Unlike the TCG specifications, a verifier in this case would need to verify the trustworthiness of the trust anchor at a given layer and build the chain of trust from the trust anchor upward. If the verifier trusts the trust anchor, then he implicitly

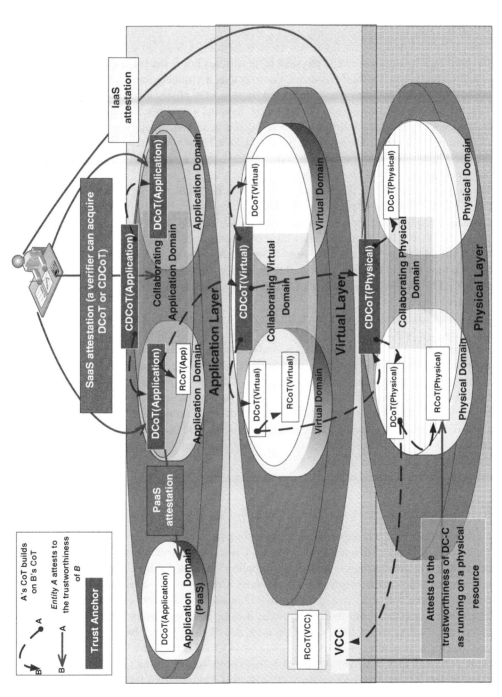

Figure 9.3 Clouds trust anchors

trusts all resources managed and attested by the trust anchor. The presented trust anchors are specific to each pool of resources at each layer, which we summarize as follows (see Figure 9.3).

- *Physical layer trust anchor*. The physical layer trust anchor is the set of physical CDCoTs which serves selected virtual resources. For example, a virtual collaborating domain composed of N virtual domains would be served by at most N physical collaborating domains. The combination of the physical collaborating domains would result in the physical layer trust anchor of the selected virtual collaborating domain. IaaS Cloud type would only need to measure the trustworthiness of the physical CDCoT that serves the virtual resources of the IaaS Cloud customer. If trusted, then the verifier can be assured that his virtual resources are served by the trusted physical resources.
- *Virtual layer trust anchor*. The virtual layer trust anchor is the serving virtual collaborating domain CDCoT. It is used mainly to provide a unified chain of trust to software application domains when building their chains of trust. The virtual CDCoT is built on a set of trusted physical CDCoTs, that is an application layer entity only needs to attest to the trustworthiness of the virtual CDCoT and does not need to worry about the set of CDCoTs at the physical layer. If the virtual CDCoT is measured to be trusted, then application layer entities which are served by the virtual collaborating domain can assume trustworthy physical and virtual management.
- *Application layer trust anchor*. The application layer has two trust anchors: the application DCoT and the application CDCoT. A verifier needs mainly to attest to the trustworthiness of the application DCoT or CDCoT. If he finds it is to be trusted, then he can be assured that the physical, virtual, and application serving components are to be trusted. If the verifier is interested in attesting to the trustworthiness of a specific application component, then he would be interested in the application DCoT. Alternatively, if the verifier is interested in accessing collaborating applications (e.g., the daisy-chain model) then the applications CDCoT is what the verifier should be seeking. A PaaS Cloud user type would typically need to attest to the DCoT of a specific software application that is supported and managed by the Cloud for the user environment.

9.6 Summary

The Cloud infrastructure is expected to be able to support Internet-scale critical applications (e.g., hospital systems and smart grid systems). Critical infrastructure services and organizations alike will not outsource their critical applications to a public Cloud without strong assurance that their requirements will be enforced. Central to this concern is that the user should be provided with evidence of the trustworthiness of the elements of the Cloud. Cloud's infrastructure complexity and dynamism make it difficult to use remote attestation mechanisms to attest to the trustworthiness of the elements of the Cloud. This chapter presents a Cloud-specific remote attestation mechanism to build Cloud trust anchors. Such trust anchors provide different types of Cloud users with appropriate levels of abstraction, which eliminate the need to get users involved in evaluating trust across Cloud complex layering. It also enables users to attest to the trustworthiness of a compositional chain of trust rather than each item of a resource chain of trust.

9.7   Exercises

Q1. What is remote attestation?

Q2. Discuss the importance of remote attestation for establishing trust in the Cloud.

Q3. What is the chain of trust?

Q4. What are the types of chains of trust in the Cloud?

Q5. Cloud users should be provided with transparent infrastructure management. Discuss the importance of the keyword 'transparent' and how it could be satisfied with the various Cloud user requirements.

Q6. Discuss how the chains of trust could help organizations when outsourcing part of their applications using IaaS in the public Cloud model.

References

[1] Trusted Computing Group. *TPM Main, Part 1, Design Principles. Specification version 1.2 Revision 103*, 2007.
[2] Trusted Computing Group. *TPM Main, Part 2, TPM Structures. Specification version 1.2 Revision 103*, 2007.
[3] Trusted Computing Group. *TPM Main, Part 3, Commands. Specification version 1.2 Revision 103*, 2007.
[4] Imad M. Abbadi. Toward trustworthy clouds' internet scale critical infrastructure. In *ISPEC '11: Proceedings of the 7th Information Security Practice and Experience Conference*, vol. 6672 of *LNCS*, pp. 73–84. Springer-Verlag: Berlin, 2011.
[5] Michael Armbrust, Armando Fox, Rean Griffith, Anthony D. Joseph, Randy H. Katz, Andrew Konwinski *et al.* Above the Clouds: A Berkeley View of Cloud Computing, 2009.
[6] Keith Jeffery and Burkhard Neidecker-Lutz. The Future of Cloud Computing – Opportunities For European Cloud Computing Beyond 2010.
[7] Sun Microsystems. Take Your Business to a Higher Level, 2009.
[8] Imad M. Abbadi and Andrew Martin. Trust in the cloud. *Information Security Technical Report*, 16(3–4):108–114, 2011.
[9] P. England, M. Peinado, and Y. Chen. An overview of NGSCB. In Chris J. Mitchell (ed.), *Trusted Computing*, pp. 115–141. IEE, 2005.
[10] S. Pearson. *Trusted Computing Platforms: TCPA technology in context*. Prentice-Hall: Englewood Cliffs, NJ, 2002.
[11] Imad M. Abbadi and Cornelius Namiluko. Dynamics of trust in clouds – challenges and research agenda. In *6th International Conference for Internet Technology and Secured Transactions (ICITST-2011)*, pp. 110–115. IEEE, December 2011.
[12] Imad M. Abbadi. Clouds trust anchors. In *11th IEEE International Conference on Trust, Security and Privacy in Computing and Communications (IEEE TrustCom-11)*. IEEE, June 2012.

10

Provenance in Clouds

Verifiers (e.g., users, forensic investigators, and even Cloud providers) should be provided with evidence about the trustworthiness of the operations management of the Cloud. Assessing the operations management of Clouds is important, but the Cloud's infrastructure complexity and dynamism make it difficult to address. This chapter establishes a framework for setting up a trustworthy provenance system. This helps in monitoring, verifying, and tracking the operations management of the Cloud infrastructure, for example it helps in the direction of proactive service management, finding the cause of incidents, customer billing assurance, security monitoring (as in the case of lessening the effects of insider threats), security and incident reporting, and tracking both management data and customer data across the infrastructural resources.

10.1 Introduction

Cloud computing is an increasingly popular approach for the processing of large data sets and computationally expensive programs. This includes scenarios that have clear requirements for maintaining the *provenance* of data, including eScience [1] and healthcare [2], where assurance in the quality and repeatability of results is essential. In addition, Clouds have their own application for provenance: the identification of the origins of faults and security violations. However, Cloud systems are structured in a fundamentally different way from other distributed systems, such as grids, and therefore present new problems for the collection of provenance data. In this chapter we discuss these problems and present a framework which helps in establishing a trustworthy provenance management system.

As discussed in Chapter 1, there are many definitions of Cloud computing which are still inconsistent. Provenance, however, is better defined. It generally refers to information that 'helps determine the derivation history of a data product, starting from its original sources' [3]. This information is clearly valuable in data-intensive computing scenarios, such as scientific computing [4], to provide assurance in the quality of results [5] and to ensure the repeatability of experiments.

We observe that the problem (if not the concept) of provenance should also be familiar to anyone involved in debugging IT systems. System administrators must identify where

Cloud Management and Security, First Edition. Imad M. Abbadi.
© 2014 John Wiley & Sons, Ltd. Published 2014 by John Wiley & Sons, Ltd.
Companion Website: www.wiley.com/go/abbadi_cloud

an error originated, what caused it, and the effects it had. This is particularly true of security violations, and provenance records are closely related to data forensics. These tasks are usually supported through *logging* and *auditing*. This is particularly difficult in complex systems with multiple layers of interacting software and hardware, such as a Cloud. Clouds are dynamic and heterogeneous by definition, and involve several components provided by different vendors which must interoperate. Tracing the origins of faults on Cloud infrastructures involves the collection of evidence and data from diverse sources with difficult-to-determine causes and effects. Cloud computing is therefore a good example of a situation where the introduction of better provenance data could provide immediate benefits for system administrators as well as users.

Logging, auditing, and historical data are of tremendous importance for establishing trust in Clouds. This data has different usage, for example proactive service delivery (incidents and security monitoring), billing, error and forensic investigation. For convenience, we refer to this data as *log records*.

Almost all Cloud resources generate this data in some way. The importance of such data and its usage is based on the following resource types. *Physical resources* generate log records related to physical resource status, security, and incident reporting. The generated data helps in the direction of finding the cause of incidents and for security monitoring. *Virtual resources* generate log records related to virtual resource status, security, and incident reporting. They also generate usage data, which is used for billing customers using IaaS Clouds. Finally, *application resources* generate log records related to application resource status, security, and incident reporting. They also generate usage data that is used for billing customers using PaaS and SaaS Clouds.

As we discussed earlier in this book, establishing trust in Cloud systems requires two mutually dependent elements: (a) support infrastructures with trustworthy mechanisms and tools to help Cloud providers automate the process of managing, maintaining, and securing their systems; and (b) methods to help Cloud users and providers establish trust in the operational management of the infrastructure. Chapter 8 focuses on both points (a and b); specifically, it establishes offline chains of trust across the distributed elements of the Cloud physical infrastructure, helping self-managed services to securely exchange management data, and it provides a mechanism enabling users to attest to the way the Cloud infrastructure is managed. The framework presented in this chapter focuses on point (a) by supporting self-managed services with a trustworthy provenance system. In addition, this chapter discusses a mechanism for establishing offline chains of trust between Cloud entities and the provenance system, collecting log records from the distributed elements of the Cloud infrastructure, associating important identification metadata with such records in a Cloud context, and securely pushing the result to the provenance system. The integrated frameworks (i.e., those presented in the previous chapter and this chapter) help in establishing trustworthy Clouds.

10.1.1   Log and Provenance

Logs and *provenance data* are distinctly different. Logs provide a sequential history of actions, usually relating to a particular process. Provenance generally refers to information that helps determine the derivation history of a data product, starting from its original sources [3]. Provenance goes beyond an individual application or process and may refer to many pieces of equipment as well as people. Throughout this chapter we refer to logs as being a source of provenance, primarily because in Cloud, logs are used in combination for a similar purpose.

Provenance is provided in Clouds through linking together log records, collected from multiple resources, to provide the complete history of an event or result. Cloud provenance, at present, is associated with the following limitations [6]: the methods followed by Clouds to support provenance queries are basic and, in many cases, such methods are developed on an ad-hoc basis by Cloud system administrators using customized scripts to address a specific event. In addition, current provenance mechanisms are object specific; that is, they do not automate the process of managing different log and audit files and linking dependent log and audit records together. Current log and audit records are not reasonably protected, which in turn affects the creditability of provenance in the Cloud. Moreover, current Cloud provenance mechanisms are deployed and fully controlled by Cloud providers; that is, Cloud users do not have control over such mechanisms, and neither can they access log and audit records. The identified limitations motivate the need to establish a trustworthy secure Cloud provenance. In the next subsection we discuss the complexities involved in this.

10.1.2   Problem Description and Objectives

We believe that establishing trustworthy secure Cloud provenance requires great effort from both academia and industry. One of the main reasons for the complexity of Cloud provenance is that it uses log records which are associated with the following issues: log records are not properly managed and are dispersed amongst the complex and distributed infrastructure of Clouds, that is, most log records are scattered around the infrastructure using unstructured and unrelated text files; and log records do not adhere to any standard format (this covers both those generated by different processes and those generated by similar processes but from different manufacturers). Also, such log records do not have semantics explaining the meaning of the items of log records.

Provenance in Clouds with the above problems is not practical considering the enormous number of applications, complex infrastructure, and huge number of users of Clouds. In addition, Cloud provenance is even more complicated than traditional enterprises considering the dynamic nature of Clouds. The dynamic nature of Clouds results in the desired properties, for example resource consolidation, resilience, scalability, and high availability. However, this dynamism results in new challenges, for example building a logical sequence of events to investigate an incident for any one application requires data from many sources, including the application itself, all logs for possible virtual resources that the application could have used, and logs of all physical resources, that virtual resources could have used. Administrators must then combine this data correctly by identifying all time intervals when an application used a specific virtual resource, all possible time intervals when these virtual resources used physical resources, and then all relevant log files from all related resources. Collecting and combining data from these resources is not easy or practical considering the potential scale of Cloud systems. These, in turn, increase insider threats in the Cloud and reduce its trustworthiness, which discourages critical infrastructures from outsourcing their resources to public Clouds.

The foundation for providing Cloud provenance requires the following key elements: (i) establishing semantics and standards of log records which enable the automated understanding of log records as generated by multiple processes; (ii) storing log records in a structured, highly available, and centralized repository which enables provenance tools to easily and quickly find log records, query them, and bind related events together; (iii) providing security measures for storing, querying, transferring, and managing log records; and (iv) establishing trust in the operation of the processes managing log records which help end-users to establish trust in Cloud provenance.

Providing trustworthy secure Cloud provenance is a complex problem that requires huge collaborative efforts. This chapter represents a foundation framework for addressing this problem. In addition, it discusses in further detail points (ii) and (iii) above. In order to clarify the overall picture and put the scheme into context, the chapter also partially discusses point (iv). In addition to points (i) and (iv), this chapter does not cover the details of the following:

- A detailed database management system design for supporting provenance application requirements.
- A detailed design of the provenance application itself.
- Policy management and enforcement (e.g., log retention policy).
- Detailed discussion about VM agents that manage provenance data inside a VM (we only outlined one aspect of this, i.e. secure storage and transfer of provenance data).
- Protecting provenance data and domain credentials once decrypted in memory.
- Key management.

10.1.3 Organization of the Chapter

The chapter is organized as follows. Section 10.2 presents motivating scenarios. Section 10.3 discusses the management of provenance data and then extracts the system requirements. Section 10.4 defines our domain architecture. Section 10.5 identifies the software services and their functions. Section 10.6 provides our framework workflow. Section 10.7 provides an informal threat analysis of the system workflow. Finally, we discuss and conclude the chapter in Sections 10.8 and 10.9.

10.2 Motivating Scenarios

We now discuss the importance of provenance in a Cloud using two simple example scenarios, as illustrated in Figure 10.1. We assume that a Cloud provider has six physical servers PS_1 to PS_6, and two physical domains L_1 and L_2. L_1 is allocated physical servers PS_1 to PS_3, and L_2 is allocated physical servers PS_4 to PS_6. We also assume that the Cloud provider hosts an application *App*. The Cloud provider creates a virtual domain VD_1 in the virtual layer to run *App*. VD_1 is initially allocated one virtual resource, VR_1, to host *App*. VD_1 is associated with a policy allowing it to scale its resources when there is an increase in demand using resources from physical domain L_1.

The first example demonstrates how a simple increase in load, and the corresponding reaction from the Cloud, can result in a loss of provenance data. Assume the load on *App* has increased dramatically, then the following steps apply:

- VD_1 responds by instantiating a new virtual resource VR_2 replicating VR_1 inside VD_1.
- Now both VR_1 and VR_2 process *App*, which are hosted using L_1 – assume that VR_1 is hosted by PS_1 and VR_2 is hosted by PS_2.

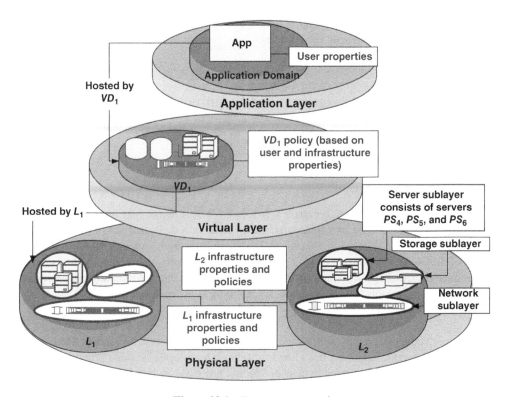

Figure 10.1   Provenance scenario

- *PS₂* has hardware problems, which results in incorrect results being generated by *App*.
- Load returns to normal and so *VD₁* downscales by removing *VR₂*.
- Cloud customers discover the problem and call the Cloud provider. If the Cloud provider only examines the log files generated by *VR₁* and *PS₁*, then they will not find the root cause of the problem or how to rectify it.

The second scenario focuses on forensic provenance in the Cloud, as follows:

- a security administrator reads the policy for *VD₁* and understands that *App* can only be hosted using *L₁* resource;
- the administrator updates the *VD₁* policy to force *VD₁* to use *L₂* resources;
- the administrator then connects to *L₂* physical resources and finds out that *VD₁* resources are running on *PS₄*, meaning that *App* is hosted there. The security administrator connects to *PS₄* and indirectly extracts important information from *App*. *PS₄* logs this activity; and
- The administrator restores the original policy, which forces *VD₁* resources to switch back to *L₁*.

If the Cloud provider only examines log files generated by *L₁* resources, then they will not discover who performed the attack or, even worse, they might never discover that an attack has happened in the first place. This is one of the main challenges that shows the importance of provenance considering the complex Cloud infrastructure and enormous distributed resources.

10.3   Log Records Management and Requirements

This section presents a provenance database scheme design and an approach for managing its records (i.e., covering point (ii) discussed in Section 10.1.2).

10.3.1   Database Design

We now discuss a provenance database scheme, a VCC scheme, and their interlinks. We categorize provenance data into four parts, each stored in a dedicated set of tables (see the provenance database scheme in Figure 10.2). Three categories cover the three horizontal layers of the Cloud taxonomy (i.e., we have physical domain provenance data, virtual domain provenance data, and application domain provenance data), while the last category represents Cloud management tools (i.e., we have management tools provenance data). Each category is composed of the following types of data:

- Log record tables having the following mandatory columns: log-id, timestamp, and detailed log data. The combination of log-id and timestamp represents the table primary key. The 'detailed log data' is generated by a layer-specific process.
- Metadata tables clarifying the relationships and interdependencies amongst different records of the 'detailed log data' in the context of Cloud taxonomy. Understanding such relationships and their interdependencies helps in automatically (and with no intervention from experts in the domain) understanding the sequential flow of events across and within the distributed elements of Clouds.

The metadata which is associated with log records varies based on the category that it represents, as follows:

- *Physical layer metadata*. This type is composed of the following columns:
  - Log-id is a foreign key binding the metadata with its set of log records, which are stored in a separate table.
  - Physical-domain-id is a unique identifier of a physical domain.
  - Physical-component-id is a unique identifier of the physical resource member of the physical domain which hosts the process generating the log records.
  - Physical-component-type identifies the nature of the physical component, that is server, storage, or network device.
- *Virtual layer metadata*. This type is composed of the following columns:
  - Virtual-domain-id is a unique identifier of a virtual domain within the virtual layer.
  - Virtual-component-id is a unique identifier of the virtual machine member of the virtual domain which hosts the process that generates the log records.
  - Physical-domain-id and physical-component-id point to the physical domain and the physical resource which host the virtual machine at the time the log record is generated.

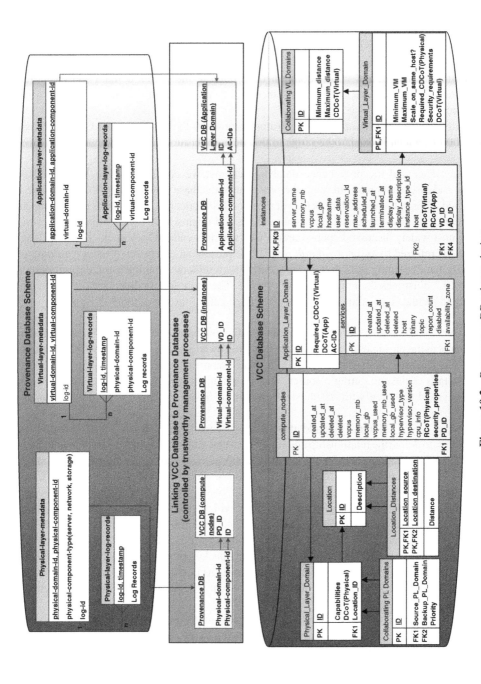

Figure 10.2 Provenance DB scheme design

- *Application layer metadata.* This type is composed of the following columns:
  - Log-id binds the metadata with its log records table.
  - Application-component-id is a unique identifier of an application component that generates the log record.
  - Application-domain-id is a unique identifier of the application domain which manages the application component.
  - Virtual-domain-id and virtual-component-id point to the virtual domain and the virtual component which runs a specific instance of the application at the time an item of log is generated.
- *Management tools metadata.* This repository represents the log records of Cloud management tools, which is outside the scope of this book to discuss.

Having discussed the provenance database scheme, we now outline the VCC database scheme and discuss its interlink with the provenance database scheme. The VCC database is composed of many tables holding details of the Cloud components. It also holds users, projects, and security details. There are many commercial and open-source VCC. This book focuses on the most widely discussed open-source VCC, which is called OpenStack. The OpenStack database is referred to as a *nova-database* [7]. We present an extension of the *nova-database* to realize the taxonomy of Clouds, user security requirements, and infrastructure properties. Figure 10.2 illustrates the *nova-database* scheme. The proposed modifications are presented using bold format. The following covers the related part of the *nova-database* scheme:

- *Compute_nodes* is an existing nova-database table that holds records reflecting a computing resource at the physical layer. We update this table by adding the following additional fields: physical resource chain of trust, RCoT(Physical); security properties that hold a list of security details of the computing resource; and a foreign key establishing the relation between the physical resource with its physical domain as exists in the Physical_Layer_Domain table.
- *Physical_Layer_Domain.* We add this table to hold the records of the Cloud physical domains, to define the relationship amongst resources, and to hold physical domain metadata. The domain metadata includes the domain capabilities, DCoT, and a foreign key pointing to the table which identifies the relative location of the physical domain within the Cloud infrastructure.
- *Collaborating_PL_Domain.* We add this table which establishes the concept of collaborating physical domains. Each record in the Collaborating_PL_Domain table identifies a specific collaborating domain (i.e., a backup domain) for each physical source domain with a priority value. A source domain can have many backup domains. The priority value identifies the order in which the physical backup domains could possibly be allocated to serve source domain needs. Backup domains are used in maintenance windows, emergencies, load balancing, etc.
- *Instances* is an existing OpenStack table representing the running instances at computing nodes. We update the table by adding the following fields: a virtual resource chain of trust RCoT(Virtual); an application resource chain of trust RCoT(Application); and two foreign keys which establish a relationship with the instance's virtual and application domain tables, as defined in the Virtual_Layer_Domain and Application_Layer_Domain tables, respectively.

OpenStack has many more tables and we also add additional tables to those discussed above, however, these are outside the scope of this book to cover (we are mainly interested in the interlinks between VCC and the provenance database). The interlink is managed by a set of trustworthy server and client software agents, which are discussed in a subsequent section (see Figure 10.3). The server software agent collects the metadata from the VCC database and pushes the result to Cloud client software agents. The metadata reflects a lively status of the elements of the Cloud infrastructure. Whenever a process wants to store a log record in the LaaS system, it sends the log record to its assigned Cloud client agent which associates the log record with the required metadata. The client agent then pushes the result to the LaaS client agent. The LaaS agent stores the result in the provenance database.

We note here that our design approach explicitly separates both databases, that is the provenance database does not interact directly with the VCC database, and the integrity of their binding data is controlled by the software agents. We follow this design approach to maintain a complete separation of management of both databases (to enforce the separation of duty principle, as our design approach assumes the provenance system is maintained by specific security administrators who do not maintain other components of the Cloud infrastructure).

10.3.2   Security Requirements

In this section we discuss the provenance system requirements and properties based on the discussed provenance database. We start with the security requirements, which are as follows:

- Provide assurance measures to the LaaS system that the log records are generated and transferred from their source by trustworthy processes.
- Provide assurance to the LaaS system that the metadata associated with each item of the log records is correct.
- Provide assurance measures to the processes which generate the log records that the Cloud management software agents are trusted to provide correct metadata and, in addition, the LaaS system is trusted to protect the log records and the associated metadata.
- Provide assurance measures to interested parties (e.g., Cloud customers, auditors, and even Cloud providers) about the trustworthiness and reliability of the LaaS system to protect the log records and associated metadata.

10.3.3   Other Requirements and Device Properties

Clouds have a huge number of resources that each host many processes. The provenance database, as a result, is expected to be highly transactional with enormous size. These properties require a careful distributed system design that maintains reliability, eliminates any single point of failure, and maintains overall high system performance. The properties of the provenance database (i.e., enormous size and high transaction rates) and the requirements of the LaaS system to be highly available and reliable with no single point of failure necessitate careful design at the infrastructure and application levels, which we outline in this subsection. LaaS, as a result, is expected to fully utilize multiple and redundant physical servers. Thus, we require the LaaS application to be installed at dedicated physical servers that do not have a virtual layer, as virtualization in the LaaS system does not add extra benefits; it rather deteriorates database

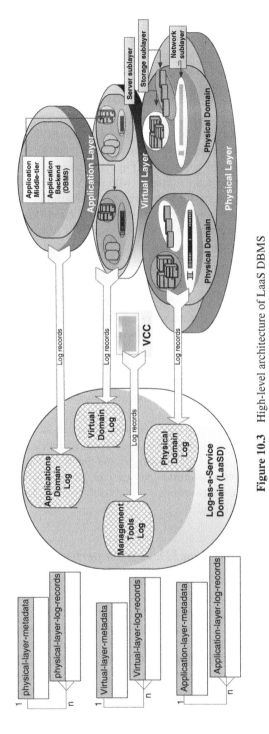

Figure 10.3 High-level architecture of LaaS DBMS

performance. The key advantages of virtualization are: consolidating resources and providing highly available and resilient design. LaaS's resources do not require consolidations as the LaaS is an isolated system that fully utilizes a bunch of distributed powerful servers. High availability and resilient design, in the LaaS system, could better be provided by different tools (using clustering technology as in the case of the Oracle Real Application Cluster (RAC)[8]). Such tools were in use even before the virtualization era. Clustering technology, in fact, provides advanced features which cannot be provided by virtualization technology at the time of writing and which are required in LaaS, for example to maintain high transaction availability [8].

We require that the LaaS system devices are commercial off-the-shelf hardware enhanced with trusted computing technology that incorporates a TPM as defined by the TCG specifications [9]. Trusted computing systems are platforms whose state can be remotely tested, and which can be trusted to store security-sensitive data in ways testable by a remote party. The TCG specifications require each TP to include an additional hardware component, the TPM, to establish trust in that platform. The TPM has protected storage and protected capabilities. The entries of a TPM PCR, where integrity measurements are stored, are used in the protected storage mechanism. This is achieved by comparing the current PCR values with the intended PCR values stored with the data object. If the two values are consistent, access is then granted and data is unsealed. Storage and retrieval are carried out by the TPM.

10.4   Framework Domain Architecture

In this section we discuss the LaaS domain architecture which forms the foundation for addressing the identified objectives. The architecture uses the dynamic domain concept as discussed in Chapter 8. The framework is composed of the following types of domain (see Figure 10.4): *log as a service domain* (LaaSD), MD, CMD, OD, and COD. The LaaSD is composed of LaaS-specific servers which host the LaaS system. Chapter 8 covers the details of MD, CMD, OD, and COD; however, the framework in this chapter extends some of the functions of the framework presented previously to make it provenance aware. Subsequent sections identify the additional functions which we introduced at the MD – this chapter does not cover integration of the provenance system with OD, COD, and CMD, as these would increase the complexity of the chapter and divert the focus.

- LaaS Domain

Definition 10.1   The LaaSD consists of platforms that host Cloud LaaS applications. Section 10.3 outlines the design requirements of the LaaSD hosting platforms. The LaaSD has a unique identifier i_{laas}, two shared unique keys k_{laas} and $k_{laas-cca}$, and a specific PKL_{laas} composed of all devices in the LaaSD. k_{laas} is used to protect log records when transferred within the LaaSD, while $k_{laas-cca}$ is used to protect log records when transferred from Cloud entities to the LaaSD (specifically between the Cloud client agent and the log client agent, as will be explained later). The credentials of the LaaSD are defined below (Definitions 10.2–10.5). The LaaSD is associated with a provenance policy, which controls its behavior and manages the provenance data, for example see the data retention policy outlined in Section 10.8.

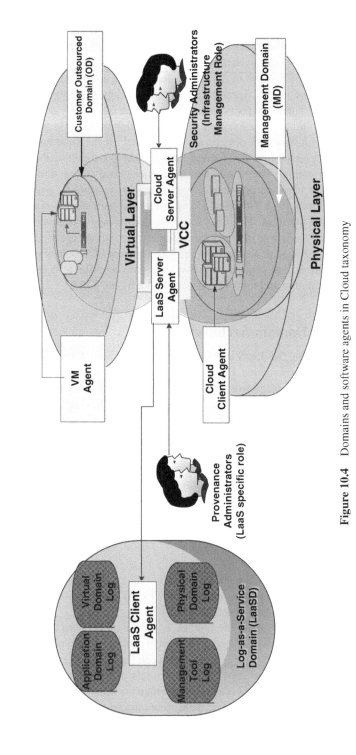

Figure 10.4 Domains and software agents in Cloud taxonomy

Definition 10.2 The LaaSD identifier i_{laas} is a unique number that we use to identify the LaaSD. It is securely generated and protected by the TPM of the VCC.

Definition 10.3 The LaaSD key k_{laas} is used to protect provenance data. k_{laas} is a symmetric key that is securely generated and protected by the TPM of the VCC. k_{laas} is not available in the clear, it is shared between all member devices of the LaaSD, and it can only be transferred from the VCC to a device when it joins the LaaSD.

Definition 10.4 The LaaSD public key list (PKL_{laas}) is an LaaSD-specific list that is composed of the public keys of all devices of the LaaSD. Provenance administrators assign devices to each LaaSD by providing each device public key to the VCC in the form of a PKL. The PKL_{laas} is securely protected and managed by the VCC.

Definition 10.5 The LaaSD key $k_{laas-cca}$ (also called the LCA-CCA key) is used to protect the provenance data when transferred from Cloud distributed elements to an LaaS provenance application. $k_{laas-cca}$ is a symmetric key that is securely generated and protected by the TPM of the VCC. $k_{laas-cca}$ is not available in the clear, it is shared between member devices of the LaaSD and MD, and it can only be transferred from the VCC to a device when the device joins an MD or LaaSD.

- Management domain

Definition 10.6 The MD is defined in Definition 8.2. However, in the extended framework the MD policy not only manages the behavior of its members and controls the behavior of collaborating MDs, but also controls the transfer of log records and the association of metadata to the LaaSD from across the distributed elements of the Cloud infrastructure. In addition, the MD credentials also include the shared LCA-CCA key provided by the LaaSD, with similar definitions to those provided above.

10.5 Framework Software Agents

The presented framework architecture is composed of a set of software agents which are required to implement the functions of the framework (see Figures 10.4 and 10.5). The software agents are as follows: Cloud client agent (CCA); Cloud server agent (CSA); LaaS server agent (LSA); LaaS client agent (LCA); and virtual machine agent (VMA). Chapter 8 provided the required protocols for the CCA and CSA, which control the management of the OD/COD at the MD/CMD. As we discussed earlier, the objectives of Chapter 8 are not the same as the objectives of this chapter, which necessitates introducing changes to the CCA and CSA to provide an integrated framework. In the remaining part of this section we discuss in

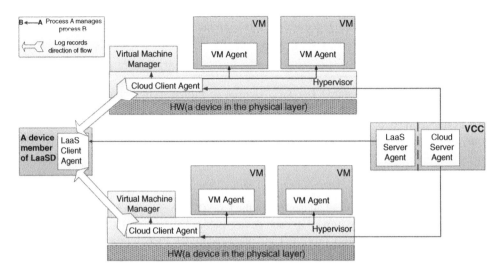

Figure 10.5   Software agents for Cloud provenance and management services

detail the functions of these agents, and the changes we introduce at the CCA and CSA to be provenance aware. The software agents still have Assumption 8.1.

10.5.1   Cloud Server Agent

The CSA is a trusted management agent that runs at the VCC, and has the following functions:

- Installs the CCA on all physical devices except those of the LaaS servers, as discussed in Chapter 8.
- Manages the MD/CMD policies, as discussed in Chapter 8. Also, manages the provenance policies. Provenance policies provide the assurance that the log records are securely generated and transferred to authorized entities. These policies also provide assurance that trustworthy metadata is generated and associated with the log records.
- Establishes offline chains of trust between Cloud entities, which include the following:
  - Chains of trust between the CSA and CCA, as covered in Chapter 8.
  - Collaborating with the LSA to establish chains of trust between the CSA and LSA, and the CCA and LCA.
- Creates and manages the MD/CMD (the creation is covered in Chapter 8).

10.5.2   LaaS Server Agent

The LSA is a trusted provenance agent which runs at the VCC. The LSA has the following functions:

- Installs and manages the LCA.
- Manages the provenance policies which provide assurance that the provenance data is only accessible to authorized entities and controls provenance data retention.

- Establishes offline chains of trust between provenance management agents (i.e., the LSA and LCA) and between provenance management agents and other agents (i.e., the LSA and CSA and the CCA and LCA).
- Creates and manages the LaaSD, which includes the following:
  - Securely generating and storing LaaSD protection keys.
  - Attesting to the execution environment status of devices' LCAs whilst being added to the domain and ensuring they are trusted to execute as expected; hence, trusted to securely store the domain key and to protect domain content.
  - Adding and removing devices to and from a domain by releasing the domain-specific key to the LCA running on devices joining the LaaSD.

10.5.3   LaaS Client Agent

The LCA is a trusted provenance agent which runs at a physical platform member of the LaaSD. The LCA has the following functions:

- Intermediates the communication between the CSA/CCA and the provenance system and between the provenance security administrators and the provenance system.
- Assures verifiers that the provenance system operates in a trusted environment; that is, it can access provenance data when its execution environment is trusted.
- Manages and enforces organization policy related to the provenance operations as distributed by the LSA.

10.5.4   VM Agent

The VM agent is a trusted agent running at all virtual machines which are organized into ODs and CODs. The VM agent intermediates the communication between running processes inside a virtual machine and the CCA – this chapter only covers the secure storage of provenance data. The VM agent attests to the execution status of all running processes inside the VM and ensures that they are trusted to behave as expected. It then securely transfers the log records to the CCA. The CCA, as explained next, is in charge of adding and binding the metadata to log records and then transferring the result to the LCA.

10.5.5   Cloud Client Agent

The CCA is a trusted client-management agent running at the resources of the physical layer (excluding members of the LaaSD). The CCA has the following functions, which are related to the provenance system (these are additional functions to those discussed in Chapter 8):

- Enforces provenance policy as distributed by the LSA via the CSA.
- Intermediates the communication between all processes running at the physical platform and the LCA. Specifically, it grabs the log records as forwarded from inside the VM and other processes in the hypervisor, and then associates them with the required metadata. Subsequently, it sends the result to its allocated LCA.
- Sends its own log records (i.e., log records related to the management of virtual resources at physical resources) to its allocated LCA.

10.6   Framework Workflow

This section discusses a possible workflow of the system framework. The chapter does not discuss OD/COD (i.e., it does not discuss VMs and the details of their agents), neither does it discuss the management of application provenance. Discussing such details would drag us into extra complexities that divert the focus of the chapter. At this early stage of the work we discuss a set of protocols as a proof of concept with an informal security analysis. This is to clarify how the framework components could possibly be managed. Once we proceed further in this work and address the identified challenges, we will then need to provide a formal analysis in which the discussed protocols would likely to be updated.

10.6.1   Cloud Server Agent Initialization

This section describes the procedure of initializing the CSA discussed in Section 10.5. The following notation is used in this section: TPM is the TPM on the VCC; S is the platform state at release as stored in the PCR inside the TPM; and (Pu, Pr) is a non-migratable key pair such that the private part of the key Pr is bound to the TPM and to the platform state S. The following protocol functions are defined in [9]: $\text{TPM}_{CreateWrapKey}$, $\text{TPM}_{LoadKey2}$, TPM_{Seal}, and TPM_{Unseal}.

The main objective of initializing the CSA is to prepare it to implement the framework of the discussed scheme. This includes the following:

- Cloud security administrators install the CSA on the VCC – the installation of the CSA includes generating a non-migratable key pair (Pr, Pu) to protect domain secrets.
- The CSA manages security administrator(s) credentials and securely stores them to be used whenever administrator(s) need to be authenticated to the CSA.

The first time security administrators run the CSA it performs the following initialization procedure (as described in Algorithm 10.1). The objective of this algorithm is to initialize the CSA. The CSA executes and sends a request to the VCC-specific TPM to generate a non-migratable key pair, which is used to protect domain secrets. The TPM then generates this key and seals it to be used by the CSA when the hosting device execution status is trusted.

The CSA then needs to ensure that only security administrators can use the CSA. For this the CSA instructs security administrators to provide their authentication credentials (e.g., password/PIN), as described in Algorithm 10.2. The objective of this algorithm is to enrol security administrators in the CSA. The CSA then requests the TPM to store the authentication credentials of the Cloud security administrators associated with its trusted execution environment state (i.e., the integrity measurement as stored in the TPM PCR) in the VCC protected storage. By storing data in a protected storage we mean 'sealing data' in TCG terms, so that data can only be accessed by the trusted server agent. The authentication credential is used to authenticate security administrators before using the CSA; see Algorithm 10.3.

Given the definitions and the assumptions above, the protocol is described by Algorithms 10.1–10.3. The objective of the protocol is to install the server agent at the VCC, which generates the non-migratable key to encrypt the CSA secrets. The protocols are used by security administrators when interacting with the server agent.

Algorithm 10.1 CSA Initialization

---

1. CSA→ TPM: TPM$_{CreateWrapKey}$.
2. TPM: Generates a non-migratable key pair (Pu, Pr).
   Pr is bound to the TPM and to the required platform state S at release, as stored in the PCR inside the TPM.
3. TPM→ CSA: TPM_KEY12[Pu, Encrypted Pr, TPM_KEY_STORAGE, tpmProof=TPM (NON-MIGRATABLE), S, Auth_data].

Algorithm 10.2 Administrators Registration

---

1. CSA→ Administrators: Request for security administrators' authentication credentials.
2. CSA→ TPM: TPM$_{LoadKey2}$(Pr).
   Loads the private key Pr in the TPM trusted environment, after verifying the current PCR value matches the one associated with Pr (i.e., S). If the PCR value does not match S, the CSA returns an appropriate error message.
3. CSA→ TPM: TPM$_{Seal}$(Authentication_Credential).

Algorithm 10.3 Authentication Verification

---

1. CSA→ Administrators: Request for authentication credentials.
2. CSA→ TPM: TPM$_{LoadKey2}$(Pr). The TPM on the CSA device loads the private key Pr in the TPM trusted environment, after verifying the current PCR value matches the one associated with Pr (i.e., S). If the PCR value does not match S, the CSA returns an appropriate error message.
3. CSA→ TPM: TPM$_{Unseal}$(Authentication_Credential).
4. TPM: Decrypts the string Authentication_Credential and passes the result to the CSA.
5. CSA: Authenticates the administrators using the recovered authentication credentials. If authentication fails, the CSA returns an appropriate error message.

10.6.2   LaaS Server Agent Initialization

The process of initializing the LSA follows exactly the same process and algorithms described for initializing the CSA in Section 10.6.1. The main differences are as follows: the LaaS should be managed by provenance security administrators who should not have access to the CSA. Similarly, the CSA security administrators should not have access to the LSA. The LaaS should have its specific non-migratable key pair, which is independent of the CSA key pair and, although both the LSA and the CSA run at the VCC this does not mean that the VCC is a single entity. It is most likely to be the opposite (as currently implemented, for example, in OpenStack), having multiple different entities that could each be allocated a specific function for scalability, performance, and security reasons.

10.6.3   LCA and CCA Initialization

This section describes the procedure of initializing client agents, which could be LCA or CCA. The goal of this procedure is to prepare devices to participate in Clouds. This covers generating a non-migratable key to protect important credentials at client devices.

The protocol of initializing an LCA and CCA is described in Algorithm 10.4. The objective of this algorithm is to install a copy of the agent, which generates a non-migratable key to protect a device's credentials. TPM, S, and (Pu, Pr) have the same meanings as earlier.

Algorithm 10.4   LCA Initialization – this applies equally to CCA initialization

---

1. LCA→ TPM: $TPM_{CreateWrapKey}$.
2. TPM: Generates a non-migratable key pair (Pu, Pr).
3. TPM→LCA: TPM_KEY12[Pu, Encrypted Pr, TPM_KEY_STORAGE, tpmProof=TPM (NON-MIGRATABLE), S, Auth_data].

10.6.4   LaaS Domain Establishment

In this section we discuss the procedure of establishing the LaaSD, which is managed by the LSA. In the provided protocol we use the same notation as described earlier. In this subsection we require that the LSA has already been installed and initialized, exactly as described in Section 10.6.2. This includes installing the LSA, which interacts with the TPM to generate a non-migratable key pair that can only be used by the agent. This key pair is used to protect LaaS secrets.

LaaSD establishment begins when provenance security administrators want to establish an LaaSD. The administrators instruct the LSA to create a new LaaSD. The server agent authenticates the administrators as described in Algorithm 10.3. If authentication succeeds, the server agent interacts with the TPM to securely generate the LaaS-specific domain key k_{laas} and identifier i_{laas}, and a specific key $k_{laas-cca}$ to be used to establish a trusted channel between the LCA and CCA. These are described in Algorithm 10.5.

On successful completion of this protocol the LaaS credentials are initialized, including the domain key, domain identifier, LCA-CCA key, and an empty PKL. These are protected by the LSA running at the VCC, which manages LaaSD membership.

Provenance security administrators assign selected physical devices to the LaaSD based on the device properties to fulfil the required overall LaaSD properties. As we discuss in Section 10.6.5, the LSA securely transfers the domain credentials to the joining log devices. It also transfers the key $k_{laas-cca}$ associated with i_{laas} to the CSA. The CSA in turn transfers the key to the joining CCA (see Section 10.6.7), which would establish an offline chain of trust between the CCA and LCAs.

Algorithm 10.5   LaaSD Establishment

---

1. LSA→ TPM: $TPM_{GetRandom}$.
    TPM generates a random number to be used as LaaSD key k_{laas}.
2. TPM→ LSA: k_{laas}.

3. LSA→ TPM: $\text{TPM}_{GetRandom}$.
   TPM generates a random number to be used as LCA-CCA key $k_{laas-cca}$.
4. TPM→LSA: $k_{laas-cca}$.
5. LSA→ TPM: $\text{TPM}_{GetRandom}$.
   LSA generates a unique number to be used as LaaSD identifier i_{laas}.
6. TPM→LSA: i_{laas}.
7. The LaaSD credentials k_{laas}, i_{laas}, $k_{laas-cca}$, and an empty PKL_{laas} are stored in the VCC
   protected storage and sealed to the LSA so that only the LSA can access these credentials
   when its execution status is trusted. This is achieved as follows:

   LSA→ TPM: $\text{TPM}_{LoadKey2}(\text{Pr})$.

   Loads the private key Pr in the TPM trusted environment to be used in the sealing
   function, after verifying the current PCR value matches the one associated with Pr (i.e., S).
   If the PCR value does not match S, the LSA returns an appropriate error message.

   LSA→ TPM: $\text{TPM}_{Seal}(k_{laas}||i_{laas}||\text{PKL}_{laas})$.

   TPM securely stores the string $k_{laas}||i_{laas}||\text{PKL}_{laas}$ using the platform protected storage,
   such that it can only be decrypted on the current platform by the LSA, and only if the
   platform runs as expected (when the platform PCR values match those associated with Pr,
   i.e., S).

10.6.5   Adding Devices to an LaaSD

This section describes the process for adding a device to an LaaSD. The following notation is
used in the provided protocol: TPM_{LCA} is the TPM of the device running the LCA; TPM_{LSA}
is the TPM of the device running the LSA; S_{LCA} is the platform state at release as stored in
the PCR inside the TPM_{LCA}; S_{LSA} is the platform state at release as stored in the PCR inside
the TPM_{LSA}; $(\text{Pu}_{LCA}, \text{Pr}_{LCA})$ is a non-migratable key pair such that the private part of the key
Pr_{LCA} is bound to TPM_{LCA} and to the platform state S_{LCA}; $(\text{Pu}_{LSA}, \text{Pr}_{LSA})$ is a non-migratable
key pair such that the private part of the key Pr_{LSA} is bound to TPM_{LSA} and to the platform
state S_{LSA}; i_{laas} is an LaaSD-specific identifier; PKL is the LaaSD public key list; k_{laas} is the
LaaSD-specific content protection key; $k_{laas-cca}$ is the LCA-CCA-specific key for protecting
content transferred between the CCA and LaaS and to establish trust between both entities;
Cert_{LSA} is the LSA device certificate; Cert_{LCA} is the joining LCA device certificate; A_{LSA} is
an identifier for the LaaS server device included in Cert_{LSA}; A_{LCA} is an identifier for the LaaS
client device included in Cert_{LCA}; $\text{Pr}_{LSA-AIK}$ is the corresponding private key of the public key
included in Cert_{LSA}; $\text{Pr}_{LCA-AIK}$ is the corresponding private key of the public key included
in Cert_{LCA}; N_1 is a randomly generated nonce; N_2 is a randomly generated nonce; $e_{\text{Pu}_{LCA}}(Y)$
denotes the asymmetric encryption of data Y using key Pu_{LCA}, and where we assume that the
encryption primitive in use provides non-malleability, as described in [10]; and SHA1 is a
one-way hash function.

The LCA sends a 'join domain' request to the LSA. This request includes the LaaSD-
specific identifier i_{laas} and is achieved as follows: LCA→LSA: Join_Domain. Two algorithms
are then initiated to add the device to the domain. The first algorithm involves the LaaS server
and client agents to mutually authenticate each other, conforming to the three-pass mutual
authentication protocol [11]. The LSA sends an attestation request to the LCA to prove its

trustworthiness; the LCA then sends the attestation outcome to the LSA. These steps are achieved using Algorithm 10.6.

Adding a device into a domain uses Algorithm 10.7, which starts upon successful completion of Algorithm 10.6. The objective of Algorithm 10.7 is to securely transfer the keys k_{Iaas} and $k_{Iaas-cca}$ to the LCA. Both keys are sealed on the device hosting the LCA, so that they are only released to the LCA when its execution environment is as expected. If the execution status of the device running the LCA is trusted, the LSA checks if the device's public key is included in the public key list of the domain. If so, it securely releases the domain-specific key k_{Iaas} and the LCA-CCA-specific key to the LCA using Algorithm 10.7. The keys are sealed on the LCA's device, so that they are only released to the LCA when its execution environment is as expected.

Upon successful completion of these algorithms the LaaS client and server agents establish a trusted secure communication channel that is used to transfer the LaaSD key and policy to the LCA. The established secure channel, importantly, provides assurance to the LSA about the state of the client agent and forces future use of the transferred key to the agent on a specific trusted state. The device hosting the LCA is now part of the domain, as it possesses a copy of the key k_{Iaas} and its public key matches the one stored in the server agent. Member devices of the domain can access the domain log records, which are now shared by all member devices of the LaaSD.

Algorithm 10.6 LCA and LSA Mutual Authentication

---

1. LSA→ TPM$_{LSA}$: TPM$_{GetRandom}$.
2. TPM$_{LSA}$ →LSA: Generates a random number to be used as a nonce N_1.
3. LSA→ TPM$_{LSA}$: TPM$_{LoadKey2}$(Pr$_{LSA-AIK}$).
   Loads the server agent hosting device AIK in the TPM trusted environment, after verifying the current PCR value matches the one associated with Pr$_{LSA-AIK}$.
4. LSA→ TPM$_{LSA}$: TPM$_{Sign}$(N_1).
5. TPM$_{LSA}$ → LSA →LCA: N_1||Cert$_{LSA}$||Sign$_{LSA}$(N_1).
6. LCA: Verifies Cert$_{LSA}$, extracts the signature verification key of the LSA from Cert$_{LSA}$, and checks that it has not been revoked, e.g. by querying an OCSP service [12]. The LCA then verifies the message signature. If the verification fails the LCA returns an appropriate error message.
7. LCA→ TPM$_{LCA}$: TPM$_{GetRandom}$.
8. TPM$_{LCA}$ →LCA: Generates a random number N_2 that is used as a nonce.
9. LCA→ TPM$_{LCA}$: TPM$_{LoadKey2}$(Pr$_{LCA-AIK}$).
   Loads the private key Pr$_{LCA-AIK}$ in the TPM trusted environment, after verifying the current PCR value matches the one associated with Pr$_{LCA-AIK}$.
10. LCA→ TPM$_{LCA}$: TPM$_{CertifyKey}$(SHA1(N_2||N_1||A_{LSA}||i_{Iaas}),Pu$_{LCA}$). TPM$_{LCA}$ attests to its execution status by generating a certificate for the key Pu$_{LCA}$.
11. TPM$_{LCA}$ →LCA: N_2||N_1||A_{LSA}||Pu$_{LCA}$||S_{LCA}||i_{Iaas}||
    Sign$_{LCA}$(N_2||N_1||A_{LSA}||i_{Iaas}||Pu$_{LCA}$||S_{LCA}).
12. LCA→LSA: N_2||N_1||A_{LSA}||Pu$_{LCA}$||S_{LCA}||i_{Iaas}||Cert$_{LCA}$||
    Sign$_{LCA}$(N_2||N_1||A_{LSA}||i_{Iaas}|| Pu$_{LCA}$||S_{LCA}).

13. The LSA verifies Cert_{LCA}, extracts the signature verification key of the LCA from the certificate, and checks that it has not been revoked, e.g. by querying an OCSP service. The LSA then verifies the message signature, the message freshness by verifying the value of N_1, and then verifies it is the intended recipient by checking the value of A_{LSA}. The LSA determines if the LCA is executing as expected by comparing the platform state given in S_{LCA} with the predicted platform integrity metric. If these validations fail, the LSA returns an appropriate error message.

Algorithm 10.7   Sealing LaaSD Credentials to the LCA

---

1. $\text{LSA} \rightarrow \text{TPM}_{LSA}$: $\text{TPM}_{LoadKey2}(\text{Pr}_{LSA})$.
   The TPM on the LSA loads the private key Pr_{LSA} in the TPM trusted environment, after verifying the current PCR value matches the one associated with Pr_{LSA} (i.e., S_{LSA}). If the PCR value does not match S_{LSA}, the server agent returns an appropriate error message.
2. $\text{LSA} \rightarrow \text{TPM}_{LSA}$: $\text{TPM}_{Unseal}(k_{laas}||k_{laas-cca}||i_{laas}||\text{PKL})$.
3. $\text{TPM}_{LSA} \rightarrow \text{LSA}$: Decrypts the string $k_{laas}||k_{laas-cca}||i_{laas}||\text{PKL}$ and passes the result to LSA.
4. The LSA verifies i_{laas} matches the recovered domain identifier and Pu_{LCA} is included in the PKL. If so, the LSA encrypts k_{laas} and $k_{laas-cca}$ using the key Pu_{LCA} as follows: $e_{\text{Pu}_{LCA}}(k_{laas}||k_{laas-cca})$.
5. $\text{LSA} \rightarrow \text{TPM}_{LSA}$: $\text{TPM}_{CertifyKey}(\text{SHA1}(N_2||A_{LCA}||e_{\text{Pu}_{LCA}}(k_{laas}||k_{laas-cca})),\text{Pu}_{LSA})$.
6. $\text{TPM}_{LSA} \rightarrow \text{LSA}$: Attests to its execution status by generating a certificate for the key Pu_{LSA} and sends the result to the LSA.
7. $\text{LSA} \rightarrow \text{LCA}$: $N_2||A_{LCA}||\text{Pu}_{LSA}||S_{LSA}||e_{\text{Pu}_{LCA}}(k_{laas}||k_{laas-cca})||$
   $\text{Sign}_M(N_2||A_{LCA}||e_{\text{Pu}_{LCA}}(k_{laas}||k_{laas-cca})|| \text{Pu}_{LSA}||S_{LSA})$.
8. The device LCA verifies the message signature, that it is the intended recipient by checking the value of A_{LCA}, and verifies the message freshness by checking the value of N_1. If these verifications succeed, the LCA stores the string $e_{\text{Pu}_{LCA}}(k_{laas}||k_{laas-cca})$ in its storage.

10.6.6   Establishing Trust between Server Agents

Before establishing an MD domain we should first establish a chain of trust between both the CSA and LSA. This would help in establishing a transparent chain of trust between the CCA running at each member device of the MD and the LCA that runs at each member device of the LaaSD, as we discuss later. For clarity, we do not assume that the LSA and CSAs are hosted at a single VCC (as indicated earlier, the VCC could be composed of multiple but collaborating entities). The following notation is used in the provided protocol: TPM_{LSA} is the TPM of the device running the LSA; TPM_{CSA} is the TPM of the device running the CSA; S_{LSA} is the platform state at release as stored in the PCR inside the TPM_{LSA}; S_{CSA} is the platform state at release as stored in the PCR inside the TPM_{CSA}; $(\text{Pu}_{LSA}, \text{Pr}_{LSA})$ is a non-migratable key pair such that the private part of the key Pr_{LSA} is bound to TPM_{LSA} and to the platform state S_{LSA}; $(\text{Pu}_{CSA}, \text{Pr}_{CSA})$ is a non-migratable key pair such that the private part of the key Pr_{CSA} is bound to TPM_{CSA} and to the platform state S_{CSA}; i_{laas} is an LaaSD-specific identifier; i_{md}

is an MD-specific identifier; k is a specific shared key between Cloud and LSAs; Cert_{CSA} is the LSA device certificate; Cert_{LSA} is the CSA device certificate; A_{CSA} is an identifier for the LSA device included in Cert_{CSA}; A_{LSA} is an identifier for the CSA device included in Cert_{LSA}; $\text{Pr}_{CSA-AIK}$ is the corresponding private key of the public key included in Cert_{CSA}; $\text{Pr}_{LSA-AIK}$ is the corresponding private key of the public key included in Cert_{LSA}; N_1 is a randomly generated nonce; N_2 is a randomly generated nonce; $e_{\text{Pu}_{LSA}}(Y)$ denotes the asymmetric encryption of data Y using key Pu_{LSA} and where we assume that the encryption primitive in use provides non-malleability, as described in [10]; and SHA1 is a one-way hash function.

The LSA sends an 'establish trusted channel' request to the CSA as follows: LSA→CSA: Establish_Trusted_Channel. Two algorithms are then initiated to establish the trusted channel and to transfer management data across. The first algorithm involves the LSA and CSA to mutually authenticate each other, conforming to the three-pass mutual authentication protocol [11]. The agents attest to each other to prove their trustworthiness. These steps are achieved using an algorithm which is exactly the same as Algorithm 10.6. The second algorithm (Algorithm 10.8) starts upon successful completion of Algorithm 10.6. The objective of this algorithm is to securely establish a shared key k that can only be accessed by both agents when their execution status is as expected. Upon successful completion of the two algorithms, the LSA and CSAs establish a trusted secure communication channel that is used to transfer the related provenance policy and other secret data between both agents. In addition, such a trusted channel, as we discuss later, would help in establishing a transparent chain of trust between LCAs and CCAs. The established trusted secure channel provides assurance to both agents about their states and forces future use of the transferred key to be on a specific trusted state. The next sections build on the successful completion of the provided protocols when storing and querying log records, and when validating the trustworthiness of the log management processes.

Algorithm 10.8 Sealing the LSA-CSA Shared Key to the LSA/CSA Server Agents

---

1. Note that, as indicated in the text, for this algorithm to make sense it must be read after the attestation algorithm to show the reader how both entities (i.e., LSA and CSA) attest to each other's execution environment and exchange certificates.
2. CSA→ TPM: $\text{TPM}_{GetRandom}$.
   TPM generates a random number to be used as a shared key k.
3. TPM→ CSA: k.
4. k is stored in the CSA protected storage and sealed to the CSA so that only the CSA can access the key when its execution status is trusted. This is achieved as follows:
   CSA→ TPM: $\text{TPM}_{LoadKey2}(\text{Pr})$.
   Loads the private key Pr in the TPM trusted environment to be used in the sealing function, after verifying the current PCR value matches the one associated with Pr (i.e., S). If the PCR value does not match S, the CSA returns an appropriate error message.
   CSA→ TPM: $\text{TPM}_{Seal}(k)$.
   The TPM securely stores the key k using the platform protected storage, such that it can only be decrypted on the current platform by the CSA, and only if the platform runs as expected (when the platform PCR values match those associated with Pr, i.e., S).
5. The CSA then encrypts k using the key Pu_{LSA} as follows: $e_{\text{Pu}_{LSA}}(k)$.

6. $CSA \rightarrow TPM_{CSA}$: $TPM_{CertifyKey}(SHA1(N_2||A_{LSA}||e_{Pu_{LSA}}(k)),Pu_{CSA})$.
7. $TPM_{CSA} \rightarrow CSA$: Attests to its execution status by generating a certificate for the key Pu_{CSA}, and sends the result to the CSA.
8. $CSA \rightarrow LSA$: $N_2||A_{LSA}||Pu_{CSA}||S_{CSA}||e_{Pu_{LSA}}(k)||$
   $Sign_{CSA}(N_2||A_{LSA}||e_{Pu_{LSA}}(k)|| Pu_{CSA}||S_{CSA})$.
9. The LSA verifies the message signature, that it is the intended recipient by checking the value of A_{LSA}, and verifies the message freshness by checking the value of N_1. If these verifications succeed, the LSA stores the string $e_{Pu_{LSA}}(k))$ in its storage.

As in the case of the CSA, the key k can only be decrypted on the current platform by the CSA, and only if the platform runs as expected.

10.6.7 MD Establishment and Management

In this subsection we require that the CSA has already been installed and initialized, the LaaSD has been established, and a trusted channel between the LSA and CSA has been established, exactly as described earlier in Sections 10.6.1, 10.6.4, and 10.6.6, respectively. The establishment of an MD follows similar steps to those provided in Algorithm 10.5, with the following changes: the CSA does not generate the shared LCA-CCA key, it rather requests it from the LSA using the trusted channel established in Algorithm 10.8; and after the CCA receives this key, it securely stores the key along with other MD credentials.

Adding a device to the MD also follows similar steps to those provided in Algorithms 10.6 and 10.7, with the following changes: the mutual authentication protocol (Algorithm 10.6) needs to be updated to establish a chain of trust between the CSA and CCA rather than the LSA and LCA; a chain of trust needs to be established between the LCA and CCA in Algorithm 10.7. This is transparently established when the CSA sends the shared LCA-CCA key to the CCA (how this is achieved is discussed in Section 10.8); the CSA regularly receives changes relating to provenance management and policies from the LSA using the trusted channel established in Algorithm 10.8; and the CSA (by collaborating with the LSA sends to the CCA the metadata to use with the log records (such as the physical device-id reflecting the CCA's device identifier at the VCC database, the MD-id the CCA is a member of the CMD-ids the MD is a member of the VMs the CCA would manage, and the policy that controls how the CCA interacts with the LCA).

10.6.8 Secure Log Storage

In this section we discuss a possible approach for storing Cloud provenance data using an LaaSD. We list the main steps for storing a log record generated by a process P which is hosted at a physical device D. Whenever a process P generates a log record, *LOG*, it sends the *LOG* to the CCA running at D as follows:

$P \rightarrow CCA$: $LOG||APP_{ID}$ (where APP_{ID} is the process unique identifier which produces the *LOG*).

The CCA, as discussed earlier, is assigned to an LaaSD and a pre-agreed shared CCA-LCA-specific key, $k_{laas-cca}$. Such a key can only be accessed by the assigned agents when their execution environment is as expected. We assume, for performance reasons, that the CCA and LCAs keep such keys pre-loaded in memory (we assumed in Assumption 8.1 that a mechanism is in place to protect sensitive data whilst being in memory). Loading such a key is done as follows:

1. CCA → TPM: $TPM_{LoadKey2}$(Pr). The TPM on D loads the private key Pr in the TPM trusted environment, after verifying the current PCR value matches the one associated with Pr (i.e., S). If the PCR value does not match S, the CCA returns an appropriate error message.
2. CCA → TPM: $TPM_{Unseal}(k_{laas-cca})$.

The CCA associates additional metadata representing the virtual and physical layer details (i.e., virtual domain id (VD_{ID}), virtual machine id (VM_{ID}), physical machine id (PH_{ID}), and physical domain id (PHD_{ID})), and then encrypts the string using the shared key $k_{laas-cca}$. The CCA then sends the result to the LaaS agent as follows:

$$CCA \rightarrow LaaS: e_{k_{laas-cca}}(LOG||APP_{ID}||VM_{ID}||VD_{ID}||PH_{ID}||PHD_{ID})$$

As discussed above, we require that the LaaS pre-loads the shared key $k_{laas-cca}$. The LaaS then decrypts the string, and re-encrypts only the LOG field using the LaaSD-specific key k_{laas} as follows:

1. LaaS → TPM: $TPM_{LoadKey2}$(Pr). The TPM on an LaaS device loads the private key Pr in the TPM trusted environment, after verifying the current PCR value matches the one associated with Pr (i.e., S). If the PCR value does not match S, the LaaS returns an appropriate error message.
2. LaaS → TPM: $TPM_{Unseal}(k_{laas}||k_{laas-cca})$.
3. LaaS decrypts the string $e_{k_{laas-cca}}(LOG||APP_{ID}||VM_{ID}||VD_{ID}||PH_{ID}||PHD_{ID})$.
4. LaaS then encrypts the LOG field as follows: $e_{k_{laas}}(LOG)$.

Finally, the LaaS stores the encrypted LOG record and the extracted metadata in a set of tables inside the provenance DBMS (identified in Section 10.3). We require that the LaaS DBMS provides additional protection measures for the stored provenance data. An example of this is the Oracle Wallet [13]. In this, the DBMS automatically stores the data encrypted inside the DBMS. It is beyond the scope of this book to discuss or analyze the process of securely storing data inside a DBMS.

10.7   Threat Analysis

In this section we informally analyze the threats, services, and mechanisms for the provenance framework workflow presented in Section 10.6. We focus on the threats, services, and mechanisms that apply to provenance and management data, and the domain credentials of MD and LaaSD.

Provenance and Cloud security administrators, when interacting with the server agents running at the VCC, could violate their privileges by adding unauthorized devices to a domain or even an unauthorized party could steal security administrators' authentication credentials to add an unauthorized device into a domain. The *administrators' authorization violation threat* can be mitigated by combining different measures, for example: requiring that N out of M administrators successfully authenticate themselves directly to the VCC for request authorization; using logging and auditing mechanisms that could detect abnormalities in the system; and using the policy of separation of duty, for example, preventing administrators (both provenance and Cloud) from accessing log files, which are routinely examined by auditors. The *stealing of administrators' credentials*, in contrast, can be mitigated by using strong authentication measures which involve a combination of: 'something the administrator has,' for example a smart card; 'the security administrator is...' for example biometric verification; and/or 'the security administrator knows...' for example a password or PIN. At this foundation stage, we do not cover the implementation and enforcement of such mechanisms.

The server software agents running at the VCC raise the following security threats when processing and storing system credentials: *unauthorized manipulation of system credentials during use in the VCC*, and/or *unauthorized manipulation of system credentials whilst stored in the VCC*. The *confidentiality and integrity protection of system credentials during execution in a VCC* requires process isolation techniques, in which software agents run in isolation, free from being observed or compromised by other processes running in the same protected partition, or by software running in any insecure partition. This chapter does not cover this point, however, we assumed in Assumption 8.1 that such a protection mechanism is in place. The *confidentiality and integrity of system credentials whilst stored in the VCC* requires protected storage capabilities, as discussed in Section 10.3.3 and Algorithm 10.5. The protected storage capabilities uses TPM functions to protect domain credentials. The TPM is tamper-evident and so it is not easy for the protected credentials to get hacked in normal circumstances. However, the TPM cannot protect itself from physical attacks and, in addition, domain keys could possibly be revealed in different ways (such as brute-force attack). Lessening the impact of such threats requires key management. The Cloud policy makers decide on the key management policy (e.g., frequency of refreshing domain keys, what should happen if a device is hacked, etc.). In this chapter we do not cover the key management part, neither do we consider policy management and enforcement mechanisms.

The interaction between a client software agent running on a device joining a domain and the corresponding server software agent running at the VCC raises the following threats to the corresponding domain key whilst in transit: *unauthorized reading or alteration of the domain key whilst in transit, the VCC wittingly/unwittingly sending the domain key to a malicious entity, a device wittingly/unwittingly receiving the domain key from a malicious entity,* and *a replay of communications between the VCC and the added device*. The *confidentiality and integrity of the domain whilst in transit*, as discussed in Section 10.6.5, is provided by the use of asymmetric encryption where we assume that the encryption primitive in use provides non-malleability. *Entity authentication of a device to a VCC* involves a protocol exchange between the device and the VCC, as discussed in Algorithm 10.6. It is initiated when the VCC and the joining device mutually authenticate each other. This mutual authentication attests to the scheme applications execution status and whether the platform is trusted. By this the VCC can only communicate with a trusted entity, and so cannot unwittingly send the domain key to a malicious entity. Similarly, the device agent, if it is not operating properly, cannot get the domain key and so it cannot wittingly send it to a malicious entity (see Algorithms 10.7 and 10.8). A similar discussion also applies to *entity authentication of a VCC to a device*. *Prevention of replay of communications between a VCC and a device* is provided by the inclusion of nonces in protocol messages (see Section 10.6.5).

Domain devices raise the following threats to the processing and storage of the domain key and content: *unauthorized reading or alteration of the domain key during use in the device*, *unauthorized reading or alteration of the domain key whilst stored in the device*, *unauthorized reading or alteration of content during use in the device*, and *unauthorized reading or alteration of content whilst stored in the device*. The *confidentiality and integrity of the domain key during execution on a device* is covered in Assumption 8.1 as discussed above for the VCC. The *confidentiality and integrity of the domain key whilst stored in a device*, as discussed above, not only requires protected storage capabilities but also key and policy management and enforcement mechanisms. The *confidentiality and integrity of domain content during execution on a device* follows the same discussion as the point of protecting the domain key during execution in the device. The *confidentiality and integrity of domain content* is protected by encrypting it using the domain key whilst stored on a device, where we assume that the encryption primitive in use provides authenticated encryption. The encryption key is bound to the device's trusted environment, as discussed in Section 10.6.8.

10.8   Discussion and Future Directions

10.8.1   Establishing Trust

In this part we discuss the foundation of trust establishment between different Clouds entities. A client or a verifier (which could, for example, be a Cloud customer, Cloud employee, or a third party) needs to assess the trustworthiness of a running application in the Cloud. This includes assessing the trustworthiness of a Cloud to manage the infrastructure and the provenance system. If the result is positive, the verifier can then trust the operation of the Cloud and would only need to assess the trustworthiness of the running application. We now discuss how the framework proceeds in this direction in more detail – it is beyond the scope of this chapter to go into the details of trust measurement and this is a planned future work.

As we discussed earlier, one of the responsibilities of the LSA is to establish a trustworthy LaaSD to manage the provenance data of Cloud elements. The first step is to install LCAs at carefully selected log-specific devices. The LSA then verifies the trustworthiness of the LCA and assures users of the trustworthy behavior of the LCA when managing the LaaSD. In other words, an untrusted LCA will automatically be evicted from managing the LaaSD. Thus, a verifier only needs to measure and then assess the trustworthiness of the LSA. If trusted, the verifier can then implicitly assume that the LCA (which is managed by the LSA) is trusted to manage the LaaSD. Assessing the trustworthiness of the LSA is not enough by itself. This is because the operation of the Cloud infrastructure (e.g., hosting of billing applications) is managed by the CSA and CCA, while the log records are managed by the LSA and LCA.

Therefore, a verifier would also need to measure and then verify the trustworthiness of the CSA as well as the LSA. As in the case of assessing the trustworthiness of log management, a verifier does not need to measure and assess the trustworthiness of the CCA. It is rather the opposite, as the verifier should not, indeed, get involved in understanding the complexities of the Cloud infrastructure [14]. As in the case of the LSA, one of the key functions of the CSA is to assure users that only trustworthy CCAs can manage the Cloud infrastructure and untrusted agents will automatically be evicted from the MD.

A chain of trust is also required between both the CCA and LSA, which is provided based on the above chains of trust, as follows: we established a chain of trust between the LSA and LCA; we established a chain of trust between the LSA and CSA; and we established a chain of trust between the CSA and CCA. Using these chains of trust, we have established a transparent chain of trust between CCAs and LCAs.

To conclude, a verifier should not (i.e., must not) get involved in understanding the details of the Cloud infrastructure. The identified chains of trust help in this direction, as a verifier only needs to attest to the trustworthiness of the requested application and the VCC which runs both the LSA and CSA.

10.8.2   Log Retention

Log retention policy is an important topic that needs to be considered in an LaaS system. It is about how long log records should be kept before deletion or offline archiving. The log retention policy would be based on the type of log records which, as discussed, we categorize as follows: physical, virtual, application, and management tools log records. The application layer log records are relatively large in comparison with the others. Provenance might require the co-existence of all these log records. A retention policy would also depend on the log nature, owner requirements, and other legislative measures. For example, EU legislation requires that log records of financial applications should be kept for at least five years.

Performance could also be an issue with such large log data if archiving is not considered. As discussed earlier, our database design is built on a distributed database management system that utilizes existing technology to manage such a large database, for example clustering technology, table partitioning, and data replication. Log records archiving would be based on insertion date as live and archive. Live records could, for example, represent the last six months of activity while archive log records represent data older than that. Live records should be stored in high-performance devices, while archiving data could be stored in slower devices. Such architectural decisions must be defined, in advance, by policy makers. We have prototyped this design approach and found that performance is enhanced dramatically, in particular live records are frequently accessed while archive records are required mainly for forensic investigation cases.

10.8.3   Achievement of Objectives

Section 10.1.2 identifies four key requirements for trustworthy secure Clouds provenance, and we now discuss those covered in this chapter. We partially addressed requirement (ii) as follows: providing a high-level design of a provenance system which is built on a distributed DBMS engine; associating each item of the log record with a metadata identifying the recorded log in the context of the Cloud taxonomy; and identifying the provenance system requirements. We covered requirement (iii) as follows: establishing an LaaSD which manages the secure sharing of provenance data between LaaSD member devices; updating the framework presented in Chapter 8 to associate provenance metadata with log records; and integrating the frameworks presented in Chapter 8 with this chapter's framework to enable the secure transfer of log records from their originating processes to the log repository. The previous subsection discusses how we partially covered point (iv), which is related to trust management – more work is still needed on this point, which is related to trust evaluation in Clouds.

Section 10.6.8 provides a possible approach to how the integrated framework could work. However, this is not enough in itself to assure provenance data integrity and confidentiality whilst being stored and processed within the LaaSD. For example, this chapter has not discussed key management, policy management, and protecting sensitive data whilst being processed. These subjects are too complex, especially in the Cloud context, to be covered in this chapter, and we leave them as planned future work.

10.9 Summary

This chapter has discussed an important topic in Clouds computing, which has not yet received much attention. It presents a framework for trustworthy Cloud provenance. Cloud provenance is a key requirement to establish a foundation for providing trust in the Cloud. Establishing trust in the Cloud requires trustworthy self-managed services that can automatically, and with minimal human intervention, manage Cloud users' resources at the Cloud infrastructure. Such self-managed services require trustworthy Cloud provenance as it helps in taking the right action on changes and incidents. Cloud provenance has many additional advantages, for example it is a key requirement in forensic investigation. This chapter does not provide an exhaustive secure framework, neither does it provide a formal security analysis of the framework. For example, the chapter does not cover key management or database security. This is because discussing such topics is a whole area of research in the Cloud context. In addition, the discussed framework still requires further extensions which, likely, would introduce some changes. The framework addresses part of the identified requirements and establishes a foundation for further research to address other requirements.

10.10 Exercises

Q1. Logs and provenance are distinctly different. How can logs be used as a source of provenance in Clouds? Is it sufficient only to use logs as a source of provenance?

Q2. Discuss the importance of provenance in different Cloud deployment types.

Q3. What are the shortcomings of the current mechanisms of Cloud provenance?

Q4. What are the key challenges for establishing trustworthy Cloud provenance?

References

[1] Rajendra Bose and James Frew. Lineage retrieval for scientific data processing: A survey. *ACM Computing Survey*, 37(1):1–28, 2005.

[2] J. Vázquez-Salceda, S. Alvarez, T. Kifor, L. Z. Varga, S. Miles, L. Moreau, and S. Willmott. In R. Annicchiarico, U. Cortés, and C. Urdiales (eds), Agent Technology and E-Health, chap. EU PROVENANCE Project: An Open Provenance Architecture for Distributed Applications. Whitestein Series in Software Agent Technologies and Autonomic Computing. Birkhäuser: Switzerland, 2007.

[3] Yogesh L. Simmhan, Beth Plale, and Dennis Gannon. A survey of data provenance in e-science. *SIGMOD Record*, 34(3):31–36, 2005.

[4] Kiran-Kumar Muniswamy-Reddy, Peter Macko, and Margo I. Seltzer. Making a cloud provenance-aware. In *TaPP '09: Proceedings of the First Workshop on the Theory and Practice of Provenance*, 2009.

[5] Yolanda Gil, Ewa Deelman, Mark Ellisman, Thomas Fahringer, Geoffrey Fox, Dennis Gannon, *et al.* Examining the challenges of scientific workflows. *IEEE Computer*, 40(12):26–34, 2007.

[6] Imad M. Abbadi and John Lyle. Challenges for provenance in cloud computing. In *3rd USENIX Workshop on the Theory and Practice of Provenance (TaPP '11)*. USENIX Association, 2011.

[7] OpenStack. OpenStack Compute – Administration Manual, 2011. http://docs.openstack.org.

[8] Oracle. Oracle Real Application Clusters (RAC), 2011. http://www.oracle.com/technetwork/database/clustering/overview/index.html.

[9] Trusted Computing Group. *TPM Main, Part 2, TPM Structures. Specification version 1.2 Revision 103*, 2007.

[10] International Organization for Standardization. *ISO/IEC 18033-2, Information technology – Security techniques — Encryption algorithms — Part 2: Asymmetric ciphers*, 2006.

[11] International Organization for Standardization. *ISO/IEC 9798-3, Information technology – Security techniques — Entity authentication – Part 3: Mechanisms using digital signature techniques*, 2nd ed, 1998.

[12] M. Myers, R. Ankney, A. Malpani – S. Galperin, and C. Adams. *X.509 Internet Public Key Infrastructure Online Certificate Status Protocol – OCSP*. RFC 2560, Internet Engineering Task Force, June 1999.

[13] Oracle. Oracle Advanced Security Administrator's Guide – Using Oracle Wallet Manager, 2011. http://docs.oracle.com/cd/B10501_01/network.920/a96573/asowalet.htm.

[14] Imad M. Abbadi and Cornelius Namiluko. Dynamics of trust in clouds – challenges and research agenda. In *6th International Conference for Internet Technology and Secured Transactions (ICITST-2011)*, pp. 110–115. IEEE, 2011.

11

Insiders

The problem of insiders in organizations is one of the most complex problems to deal with. This is because insiders have to be trusted to perform their daily business processes. In Cloud, the problem of insiders is even more complicated, as the domain of insiders is bigger than organizations and insiders have higher motivation to attack Cloud applications. This chapter analyzes this problem and provides a systematic method to identify potential and malicious insiders in the Cloud environment.

11.1 Introduction

The insider problem is cited as the most serious security problem and the most difficult problem to deal with [1, 2]. As discussed by Alawneh and Abbadi [3], the insider problem in organizations is caused mainly by the holders of authorized credentials who are typically the internal and authorized employees. Such employees should successfully pass several security checks before being employed by an organization. Also, such employees have a direct contract with the organization and the organization to a certain level trusts them (e.g., based on prior experience).

In a Cloud computing context, the insider problem is more significant than in traditional organizations for the following reasons: (1) the insider domain has expanded from the organization's internal employees and contractors to also include the Cloud internal employees and contractors, Cloud customers, and Cloud third-party suppliers; (2) the organization does not have a direct relationship with the Cloud employees and cannot anticipate their level of trust; (3) other Cloud customers, who could be competitors, might share the same physical server as the organization (raising issues of multitenant architecture [4]); and (4) Clouds-of-Clouds, in which a provider might host part of their customers' data at another provider, result in expanding the customer insiders to include all the Cloud-of-Clouds insiders. All these factors increase the insider threat to an organization's sensitive assets when moving to a Cloud infrastructure.

This chapter provides a systematic method for identifying insiders. The chapter is organized as follows. Section 11.2 presents the definition of insider and potential insider. Section 11.3

Cloud Management and Security, First Edition. Imad M. Abbadi.
© 2014 John Wiley & Sons, Ltd. Published 2014 by John Wiley & Sons, Ltd.
Companion Website: www.wiley.com/go/abbadi_cloud

provides a set of models illustrating the relationship between actors, credentials, and infrastructure in the Cloud computing context, focusing on IaaS. It then provides a method for identifying insiders. Finally, Section 11.4 summarizes the chapter.

11.2   Insiders Definition

11.2.1   Background

Organizations have always faced challenges in protecting sensitive information from being revealed to unauthorized parties. Such challenges are even harder to address when an organization needs the sensitive content to be shared between its own employees to achieve a particular task. Threats from outside the organization are always a major worry; however, organizations cannot ignore the risk of authorized employees, or 'insiders,' revealing sensitive content to unauthorized parties. This is because insiders may have privileges and know where to obtain sensitive information from within the organization. Thus, the risk associated with insider threats is often greater than the risk of threats originating outside the organization [2, 5].

It is important to understand the meaning of the term 'insider' in order to correctly define the threats to an organization and determine how to tackle them. This section is dedicated to discussing the definition of the term *insider*. One of the most important criteria for insiders is that users should have authorized credentials enabling them to access an organization's sensitive content to be considered as insiders. For an attacker to be an insider he should possess authorized credentials. We therefore raise the following questions: What if an employee shares the credentials with an unauthorized party? What are the insider threats that could result from such an action? What are the impacts of such threats on an organization's sensitive content? All these questions motivate the focus on the threats that could result from authorized users when sharing their credentials.

Organizations often manage the sharing and protection of content by controlling what authorized employees are allowed to do with sensitive content. This is typically achieved by defining access rights that restrict what authorized employees can or cannot do with content. Widely discussed access control schemes such as Discretionary Access Control (DAC) [6] and Mandatory Access Control (MAC) [7] protect content by enforcing access rights where content is stored. However, when content is physically copied to another device, the access rights are not copied with the content; this means that the access rights are no longer enforceable on the other device. Such schemes are good enough to satisfy content protection requirements where content is not transferred between users [5]. However, with conventional access control systems, once content leaves the device (where it is stored) it becomes disassociated from its access rights and thus loses its protection. Thus, when insiders copy content and leak it to a third party the content access rights will no longer be enforceable on the transferred content.

Other access control models such as Role-Based Access Control (RBAC) [8] and Usage CONtrol (UCON) [9] also suffer in this respect, since they are general-purpose models and provide no explicit framework or guidance for secure sharing.

The demand for distributing digital content between an organization's departments and the need for content to be shared between employees without affecting its protection motivates the need for schemes which are capable of protecting content even after it has been distributed or shared. These schemes provide a pervasive access control policy, sometimes known as a 'sticky policy.' Enterprise rights management schemes (ERM) [10] are examples of these, which attempt to expand the policy enforcement point to cover not only where content is stored, but also where it is subsequently sent and used. Although there are major differences between traditional access control schemes and ERM schemes, they both aim to protect content from unauthorized usage. Within any organization, traditional access control schemes and ERM schemes rely on an employee's credentials to authenticate users and provide them with proper authorization rights [10]. But what if authorized employees share their credentials with an unauthorized party? Are ERM schemes still capable of protecting an organization's sensitive content? As has been discussed by Alawneh and Abbadi [3], ERM schemes cannot protect sensitive content from insiders.

11.2.2   Definition

Within the academic community there are a diversity of definitions of the term 'insider' [11]. We use the analysis provided by Alawneh and Abbadi [3], which presents a detailed analysis of insiders in organizations.

The summary of the Dagstuhl Seminar on countering insider threats [1] proposes several definitions of an insider, as follows: Someone defined with respect to a resource, leading to degrees of 'insiderness.' Somebody with legitimate, past or present, access to resources. A wholly or partially trusted subject. A system user who can misuse privileges.

So, someone with authorized access who might attempt unauthorized removal or sabotage of critical assets or who could aid outsiders in doing so.

While the above definitions may be adequate for a simple self-contained organization, they have some shortcomings when considering more complex modern enterprises. Moreover, there are more complicated cases for insiders which may not be clear in the previous definition. For example, the case of an authorized employee sharing his credentials with an unauthorized individual. The unauthorized individual may subsequently use the credentials to leak sensitive information from the employee's organization. In this case, it is not clear whether this is an 'insider' attack or an external 'masquerade' attack. Also, who is considered to be the insider, the authorized employee or the unauthorized individual?

Another, more general, definition of an insider is *someone with access, privilege, or knowledge of information systems and services* [11]. This definition considers both external and internal users who possess authorized access or knowledge as insiders. However, this definition is perhaps too general, as it does not explicitly state the conditions under which the access or knowledge was obtained.

From the above we conclude that the previous definitions of insiders need to be refined to cover more complicated insider cases. To set the insider definition we start by specifying the first time a user becomes an insider. In other words, we specify what makes an ordinary user become an insider. In real life, when an organization requires someone to work on sensitive information, it may carry out a background check on the user. The extent of this will depend on how sensitive the information is. The individual may then have to sign some statement agreeing to behave in a trustworthy manner. The organization would provide the user with credentials enabling them to access the information required for them to carry out their agreed duties. This is the point when an organization should consider the user as an insider. It is our contention that *having valid credentials* is an important requirement when considering whether a user is an insider. This requirement needs to be part of the insider definition.

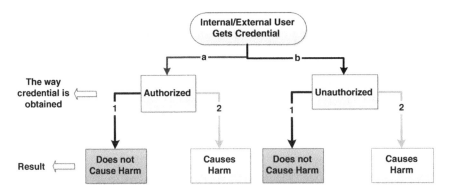

Figure 11.1   Factors affecting the definition of an insider

To extend the previous definitions of an 'insider' we consider four main factors. Figure 11.1 illustrates these in a conceptual diagram, which has the following main elements:

- A user's relation with the organization whilst performing an action on the sensitive information or content. The user could be either internal or external to the organization.
- The method used to obtain credentials. This could be either authorized or unauthorized. 'Authorized' means that credentials are granted to the user in an authorized way. 'Unauthorized' means that the user obtains the credentials either by mistake (e.g., overheard accidentally or sent by mistake) or deliberately (e.g., stolen from somewhere or obtained by social engineering).
- Result of a user's access to the organization's sensitive content. By this we mean the consequences of the user accessing content with the obtained credentials. This could be either to cause harm or not.
- The user intention when carrying out the action on content. By intention we mean either accidental or deliberate action on the content. We do not consider this factor in the insider definition as a 'deciding factor' because we suppose that a reliable system should protect itself from both accidental and deliberate actions. However, this factor is considered as a deciding factor in other research areas, for example forensic investigations.

Figure 11.1 illustrates four possible cases to consider:

- *Case I: route from a to 1.* In this case a user is (a) granted a credential in an authorized way and (1) when accessing content does not cause harm. An example of this is internal employees who are granted credentials to perform their duties. This case applies when employees do their job as expected.
- *Case II: route from a to 2.* In this case a user is (a) granted a credential in an authorized way and (2) uses it for a purpose other than the one for which it was originally intended, which results in harm. For example, internal employees who are granted credentials to perform their daily activities but, when accessing content, intentionally or accidentally misuse their privileges (e.g., updating someone's salary, deleting an important file, or leaking content to a competitor).

- *Case III: route from b to 1*. In this case a user has (b) obtained a credential in an unauthorized way but (1) does not cause harm. For example, when a user obtains some credentials by mistake (e.g., sent by mistake) but does not act on this knowledge.
- *Case IV: route from b to 2*. In this case a user has (b) obtained a credential in an unauthorized way and (2) when accessing content, caused harm. For example, an unauthorized user obtains credentials from a friend and uses this to update someone's salary, delete an important file, or leak content to a competitor.

After analyzing the above cases, Alawneh and Abbadi conclude that when dealing with insiders we need to distinguish between potential and malicious insiders, defined as follows:

Definition 11.1 A potential insider is a user who is granted a credential in an authorized way to access sensitive corporate information for a specific purpose defined by the organization (does not cause harm), or a user who obtains a credential in an unauthorized way but does not use it to cause harm.

Definition 11.2 A malicious insider (or simply an insider) is an internal or external user who 'uses credentials' obtained by either authorized or unauthorized means, to access sensitive corporate information that results in harm to the organization. Such a misuse could be either accidental or deliberate.

Based on these definitions, a 'malicious insider' could be one of the following:

- An internal employee who possesses a valid credential to access sensitive content. This access accidentally or deliberately results in harm to the organization.
- An external user who possesses valid credentials to access sensitive content. This access accidentally or deliberately results in harm to the organization. This case includes examples like contractors, third-party vendors who have access to corporate internal information, and employees from collaborating organizations who have access to each other's organization's information.
- An internal employee who obtained a valid credential, by unauthorized means, to access sensitive content and cause harm.
- An external user who obtained a valid credential, by unauthorized means, to access sensitive content and cause harm. This case includes anyone who has gained access to internal resources by masquerading as an authorized internal employee.

11.2.3   Rules of Identifying Insiders

Based on the definition, we conclude the following set of mandatory rules to identify an insider:

> R1: The insider could be either a potential insider, as defined above, or someone who managed to obtain the potential insider credentials in some way; and

R2: Uses the credential to access a resource for a purpose different from the one which the credentials were originally granted for; and

R3: This misuse results in harm to the resource owner/manager.

In the remaining part of this chapter we use these rules to identify insiders in the Cloud.

11.3   Conceptual Models

A Cloud computing-based system typically involves a number of actors, from different organizations, which interact with the system. In order to identify which actors are potential insiders and the threats emanating from their activities, we first need to identify all actors within such a system. We then need to understand the relationships among the actors, as well as their level of access to resources and assets that are part of the Cloud. To this end, we build various conceptual models that illustrate the explicit relationships among the various entities and expose any implied relationships as well as interactions between the actors and the system.

11.3.1   Organizational View

A Cloud computing-based system of IaaS type may be designed to serve the needs of different communities, including: users within a single organization or collaborating organizations (e.g., a private Cloud within an enterprise); users within a research community comprising, for example, a virtual organization; or a public community (e.g., a mixture of enterprise and individuals). In most cases, several organizations, in different capacities, will be involved in a Cloud-based system. We develop an organizational model of such a system as shown in Figure 11.2, in which we identify a generic entity *organization* as the parent of any organizational entity within a Cloud-based system.

At a minimum, a Cloud-based system comprises a *Cloud service provider*, a *Cloud customer*, and quite often a *contractor*, such as a cleaning company or hardware suppliers, that may *work_for* either a *Cloud service provider* or a *Cloud customer*. A *Cloud customer has* some *object*, such as a computation or data, that they wish to take to the Cloud. To do this, the *Cloud*

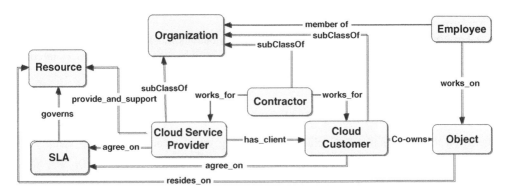

Figure 11.2   High-level organizational view of the Cloud

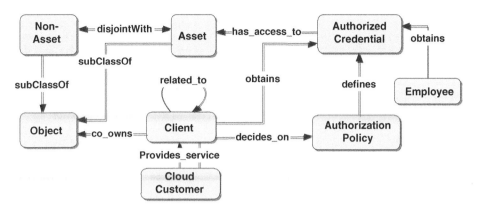

Figure 11.3   A view around the client

service provider and the *Cloud customer agree_on* some SLA which defines the *resource* provided to the *Cloud customer* and the conditions, such as performance, up time, and liability, under which the resources are provided. After this, the *object* is transferred to the Cloud and *resides_on* the resource. Furthermore, an *organization has* one or more *employees* who *work_on* the *object* owned by the Cloud customer.

11.3.2   Assets and Clients

One of the main advantages of a Cloud-based system is that the *object* can be made more accessible via the Internet to a wider audience. It therefore becomes necessary to define the entities that may have access to an object in the Cloud, as illustrated in Figure 11.3. From the organizational perspective, an object can be either an *asset* (with value to the organization) or not an asset (less valuable). These objects, especially the valuable ones, will require an *authorization policy* to define a type of *credential*, that is an *authorized credential* that enables access to the objects.

In some cases, the *object* may be co-owned by the Cloud customer and the *client* of the Cloud customers. In such cases, the *client* may have to *decide_on* all or part of the *authorization policy* to define which other clients, and sometimes employees of the Cloud customers, have access to the *authorized credential*. We clarify this in the context of an example when discussing the home healthcare system in Chapter 13.

11.3.3   Infrastructure Model

We now develop a conceptual model of a *resource*, as shown in Figure 11.4. In this model, a *resource* is a composition of components including a *physical device*, *hypervisor* (which includes a VMM), VMs, and *applications*. A component may further be divided into layers which indicate the parts of a component that interact to provide the functionality of the component. The VMM runs on top of the *physical device* to enable one or more VMs to run on the *physical device* (based on the *physical device*'s layer, as explained in Section 13.3.2). *Applications* are configured to run in VMs.

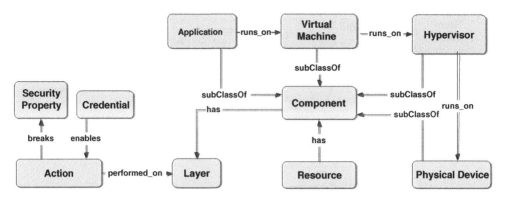

Figure 11.4   A breakdown of Cloud resources

We define an *action* as an event performed by the user of the resource. An *action* is *performed_on* a layer, and may *break* zero or more *security properties* and require some form of *credential* in order to be performed.

11.3.4   Procedure for Identifying Potential and Malicious Insiders

The conceptual models defined in earlier subsections are used for identifying potential insiders and insiders that may exist within a given context. This is achieved by instantiating the

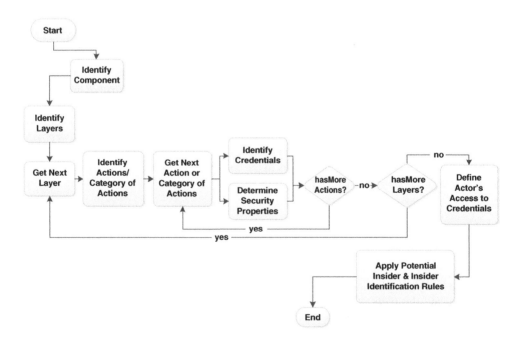

Figure 11.5   Potential insiders and insiders identification process

models given above with the actual descriptions of entities that exist within the context. More specifically, one has to provide: *layers of each of the components*; *actions that may be performed at each layer* (these should be limited to those that may have effects on one or more security properties); *credentials that may be used within the system*; and *actors that may have access to the identified credentials*.

The identification process, illustrated in Figure 11.5, involves mapping actions or categories of actions to layers on which a particular action or category of actions can be performed. Then, for each identified action or category of actions, determine the security properties that it may break and the credentials required to enable the action. With the credentials identified, identify actors that may have access to each of the identified credentials. Potential insiders and insiders are those actors that satisfy the criteria as defined in Section 11.2. This process enables us to make explicit the means through which potential insiders and insiders are defined. We use this method to identify potential insiders and insiders in a home healthcare system in Chapter 13.

11.4 Summary

This chapter has presented a set of conceptual models, which help in identifying malicious and potential insiders in a Cloud computing environment. We discuss several insider definitions and use the ones proposed by Alawneh and Abbadi [3]. Subsequently, we identified and presented the rules for identifying insiders and potential insiders.

11.5 Exercises

Q1. What is an insider?

Q2. Who are the insiders?

Q3. Discuss the importance of addressing the insider problems in Cloud computing.

Q4. Identify the insiders in a university domain when outsourcing their services to public Cloud computing.

References

[1] Matt Bishop, Dieter Gollmann, Jeffrey Hunker, and Christian W. Probst. Countering insider threats. In *Dagstuhl Seminar Proceedings 08302*, pp. 1–18. RAND Corp., Santa Monica, CA, 2008.
[2] Robert Richardson. The 12th Annual Computer Crime and Security Survey, 2007. http://i.cmpnet.com/v2.gocsi.com/pdf/CSISurvey2007.pdf.
[3] Muntaha Alawneh and Imad M. Abbadi. Defining and analyzing insiders and their threats in organizations. In *2011 IEEE International Workshop on Security and Privacy in Internet of Things (IEEE SPIoT 2011)*. IEEE, November 2011.
[4] Thomas Ristenpart, Eran Tromer, Hovav Shacham, and Stefan Savage. Hey, you, get off of my cloud: Exploring information leakage in third-party compute clouds. In *Proceedings of the 16th ACM Conference on Computer and Communications Security, CCS '09*, pp. 199–212. ACM: New York, 2009.
[5] Ravi S. Sandhu, Xinwen Zhang, Kumar Ranganathan, and Michael J. Covington. Client-side access control enforcement using trusted computing and Pei models. *Journal of High Speed Networks*, 15(3):229–245, 2006.
[6] G. Scott Graham and Peter J. Denning. Protection: Principles and practice. In *AFIPS '72 (Spring): Proceedings of the May 16–18, 1972, Spring Joint Computer Conference*, pp. 417–429. ACM: New York, 1972.

[7] Ravi S. Sandhu. Lattice-based access control models. *Computer*, 26(11):9–19, 1993.

[8] R. Sandhu, E. Coyne, H. Feinstein, and C. Youman. Role-based access control modles. In *IEEE Computer*, pp. 38–47. IEEE, 1996.

[9] Jaehong Park and Ravi Sandhu. The UCON ABC usage control model. *ACM Transactions on Information Systems and Security*, 7(1):128–174, 2004.

[10] Microsoft Corporation. Microsoft Windows Rights Management Services, 2005. http://download.microsoft. com/download/8/d/9/8d9dbf4a3b0d4ea1905b92c57086910b/RMSTechOverview.doc.

[11] Glenn Bruns, Daniel S Dantas, and Michael Huth. A simple and expressive semantic framework for policy composition in access control. In *FMSE '07: Proceedings of the 2007 ACM Workshop on Formal Methods in Security Engineering*, pp. 12–21. ACM: New York, 2007.

Part Three
Practical Examples

12

Real-Life Examples

This chapter outlines the main features of two of the most widely discussed Cloud management platforms: Amazon AWS (a commercial Cloud management platform) and OpenStack (an open-source Cloud management platform). Subsequently, the chapter presents a practical illustration of some of the concepts which are discussed throughout this book using OpenStack.

12.1 OpenStack

This section provides a high-level introduction to OpenStack.

12.1.1 What is OpenStack?

Chapter 3 presents an abstract view of Cloud management platforms, which we refer to as VCC. This section presents OpenStack [1] which is an example implementation of the VCC. OpenStack is an open-source software that was funded in October 2010 by Rackspace [2] and NASA [3]. Initially, OpenStack started by combining source codes from both RackSpace and NASA. A few months later, commercial companies started joining the OpenStack initiative. Currently, thousands of professionals around the word participate in the OpenStack software architecture and code development process [1].

OpenStack's main objective is to establish a Cloud management platform that is capable of meeting the needs of the next generation of Cloud computing. OpenStack is designed to be the global Cloud trusted management platform, and it is not meant to replace a VMM or a hypervisor function. It does not even have an implementation of a hypervisor or a VMM, and rather implements a set of APIs that interact with different hypervisors running at the Cloud's physical servers. OpenStack is designed with the objective of being an independent hypervisor, by supporting a wide range of hypervisors. This would help in stopping vendor lock-in, which is a key requirement for the future success of Clouds, as discussed in

Cloud Management and Security, First Edition. Imad M. Abbadi.
© 2014 John Wiley & Sons, Ltd. Published 2014 by John Wiley & Sons, Ltd.
Companion Website: www.wiley.com/go/abbadi_cloud

Release name	Release date	OpenStack Compute version number	OpenStack Object Storage version number
Essex	April 2012	2012.1	1.4.8
Diablo	October 2011	2011.3	1.4.3
Cactus	April 2011	2011.2	1.3.0
Bexar	March 2011	2011.1	1.2.0
Austin	October 2010	0.9.0	1.0.0

Figure 12.1   Sample of OpenStack releases – a new release is added every six-months

Chapter 1. OpenStack currently supports the following hypervisors: KVM, QEMU, ESX, ESXi, and Xen.[1]

OpenStack is still under continuous development and is still missing many components such as logging, billing, and policy management. As a result, OpenStack plans releases on a six-month basis. Importantly, its components are still immature and do not cover many aspects of the capabilities of a potential Cloud. Such limitations make OpenStack, at the current time, not suitable for a production environment (author's opinion). The main reason for the current limitations of Clouds is not related to OpenStack itself; it is rather related to the complexity and challenges of the potential Clouds which OpenStack aims to manage. OpenStack is supported by leading companies in their fields, such as IBM, vmware, Redhat, and many more.[2] Figure 12.1 lists the current releases of OpenStack.

12.1.2   Openstack Structure

In this section we briefly present the structure of OpenStack, as illustrated in Figure 12.2. It is composed of the following main components: identity service (also known as Keystone), compute service (also known as Nova), object storage service (also known as Swift), image service (also known as Glance), and dashboard service (also known as Horizon). More components are planned to be added in the future. This section briefly discusses these components.

- *Keystone*. Keystone aims to provide an identity management service when interacting with OpenStack. It covers the following functions: service management, user management, tenant management, and role management. Here, a tenant represents a group of resources (networks, volumes, instances, images, and keys) which belong to one project; a role represents an association of privileges, objects, and users.

  Keystone manages access rights using RBAC [4], which grants access rights on a tenant's objects to users. Privileges are assigned to roles in service-specific files, which are currently stored at /etc/[SERVICE_CODENAME]/policy.json. The identity service associates users with roles for tenants.

---

[1] More hypervisors are planned to be added in the future; for an up-to-date list, see http://wiki.openstack.org/ HypervisorSupportMatrix.

[2] A full list of OpenStack partners can be found at: http://www.openstack.org/foundation/companies/.

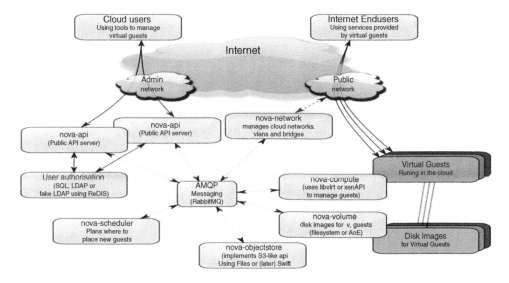

Figure 12.2   Current OpenStack structure. *Source:* http://docs.openstack.org/

- *Nova.* Nova[3] runs on physical servers to manage the resources which are allocated to servers. The resources which Nova manages include: networks, VMs, and volumes. The management of VMs covers interactions with the VMM to create, start, stop, and migrate VMs. Nova does not provide any virtualization capabilities by itself; instead, it uses an open-source library (e.g., libvirt APIs) to interact with the hypervisors as discussed previously.

  Nova has the following additional components:
  - Nova-api provides Cloud users with APIs to manage their Cloud resources.
  - Nova-database is the central repository for OpenStack data (e.g., it stores compute resources, available volumes, and instances). More databases have been added recently, for example Glance and Keystone.
  - Nova-schedule manages the hosting of VMs at the distributed servers of the Cloud infrastructure. Nova-scheduler allocates VMs to run on a physical computing node. The allocation of the physical computing node is based on the selected scheduling algorithm. Current supported schedulers include the following (see Figure 12.3):
    ○ *Simple scheduler.* This allocates VMs on physical servers by considering the least loaded host.
    ○ *Chance scheduler.* This allocates VMs randomly on physical servers.
    ○ *Filter scheduler.* This is a customized allocation of VMs on physical servers.
  - Nova-network manages the network configurations of a VM. Nova-network supports the following types – in the current release of Openstack, only one type can be configured at a time:
    ○ *Flat network.* In this a VM is allocated a fixed IP address. The IP gets injected into the VM instance on launch.

---

[3] Also called Nova-compute.

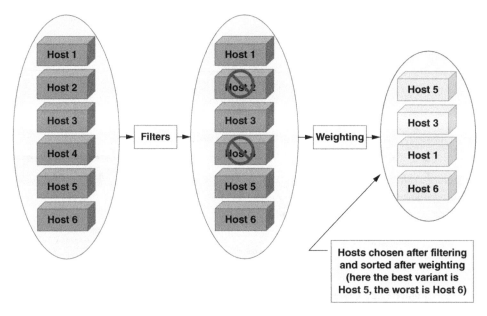

Figure 12.3   Current OpenStack scheduler. *Source:* http://docs.openstack.org/

○ *DHCP network.* In this a VM IP address is acquired by the VM instance from a DHCP server running on the nova-network.
○ *VLAN network.* In this a switch is required to support VLAN tagging.

OpenStack configures bridges and virtual interfaces. Computing nodes would typically have two interfaces: public and internal interfaces. VM traffic to the outer world (i.e., between the internal interface and the public interface) is routed via the nova-network. The allocation of an IP address to a VM could either be a static IP (fixed for the instance lifetime) or a dynamic IP (that changes dynamically).

– Nova-volume manages the VM volumes (create/delete/attach/detach). Volumes provide persistent storage for use by instances.
– Message queue is a central hub for passing actions between OpenStack components.
• *Storage components.* OpenStack supports two types of storage function: Swift and Nova-volume. Swift manages the storage of objects and provides scalable, reliable, and redundant backend storage. Swift is the storage option to consider when scalability and redundancy are required, but performance is not of concern. For example, backup and archival systems could be stored on Swift. On the contrary, Swift is not the right option for continuously accessed data as in the case of an active database management system. Swift files are exposed through an HTTP interface.

Swift has the following components, as illustrated in Figure 12.4:

– *Swift proxy.* This intermediates the communication between Cloud users and their storage.
– *Swift object server.* This stores, retrieves, and deletes the Cloud user data in the form of binary large objects (blobs).
– *Swift container server.* This groups objects in containers (similar to the directory concept but without nesting).
– *Swift account server.* This is used to list containers.

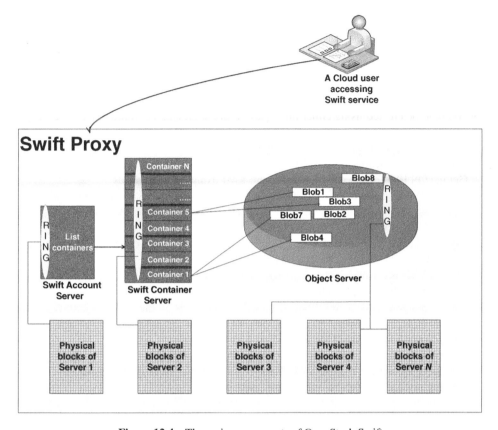

Figure 12.4   The main components of OpenStack Swift

– *RING*. This maps object names to their physical locations. Blobs, containers, and accounts have their own separate rings.

Nova-volume (or simple volume) is a detachable block storage device. The data stored on a volume is persistent even after instance deletion. A volume can be associated with only one VM instance at a time. Sharing file systems across instances should be provided using other mechanisms, such as NFS [5]. Nova-volume provides a block storage file system which is exposed through a low-level bus interface such as SCSI. Clients access the storage when the storage devices are mounted on their virtual machines.

The Nova-volume service works as follows:

– A volume group named nova-volume should first be created using the command line 'nova volume-create,' which creates a logical volume in the nova-volume group.
– After successful execution of the previous command, the compute node, which holds the volume group, would have a new logical volume as local storage. This logical volume can now be attached to a specific instance running on the compute node.
– The logical volume can be created and attached to an instance using the following commands:
  ◦ nova volume -create -display_name VOLUME_NAME SIZE_GB
  ◦ nova volume-attach INSTANCE_ID VOLUME_ID

    ◦ DESTINATION_DEVICE_MOUNT_POINT

    ◦ Logical volumes could be configured as bootable volumes.

- *Glance component*. Glance is an OpenStack image service which could be used to store and retrieve VM images (VMIs). Nova fetches the image to the hosting server and then boots it up on its host physical server. A VMI could also be managed by Swift, in which Nova follows similar processes when fetching the image to be booted up on its host physical server. VMI could be created using either third-party tools (e.g., Oz, VMBuilder, and VeeWee) or manually as follows:

  – Create an empty image using the command

    kvm-img create -f raw/qcow IMAGE_NAME SIZE

  – Get the OS ISO format and install it in the VMI using the commands

    kvm -m SIZE -cdrom ISO_PATH drive file=IMAGE_NAME,if=virtio,index=0 -boot d -net nic -net user -nographic -vnc VNCDISPLAY -monitor unix:MONITOR_FILE _NAME,server,nowait

  – Connect to the VMI instance using the VNC client and proceed with the installation instructions.

    Once a VMI is created, it can be managed as follows:

  – When using glance, upload the image to glance as follows:

    glance image-create name="NAME_IN_GLANCE" is_public=true container_format =ovf disk_format=raw/qcow2 < IMAGE_NAME

  – 'nova image-list' is one of the commands that lists images.

  – A key pair could be created for accessing an image as follows:

    nova keypair-add KEY_HANDLE_NAME > PUBLIC_KEY_FILE_NAME

  – Booting nova image can be done as follows:

    nova boot –image IMAGE_NAME –flavor m1.small –key_name KEY_HANDLE _NAME INSTANCE_NAME

  – Deleting an instance can be done as follows:

    nova delete INSTANCE_NAME

  – Pausing/suspending instances can be done as follows:

    nova pause/unpause INSTANCE_NAME_ID

  – Suspend/resume instances can be done as follows:

    nova suspend/resume INSTANCE_NAME_ID

12.1.3 Security in OpenStack

As discussed earlier, OpenStack supports identity management using the Keystone service. It also manages access rights using the RBAC mechanism. OpenStack uses the concept of security groups to provide inbound network traffic filtering for instances. A security group is a collection of rules in an IP table that gets applied to the incoming packets of instances. Each security group can have multiple rules associated with it. Each rule specifies the source IP, protocol type, destination ports, etc. Only packets matching the rule are allowed in. A security group that does not have any rules associated with it causes blocking of all incoming traffic. A security group is attached to an instance on start-up. Outbound network traffic filtering, in contrast, needs to be implemented from inside VM instances.

OpenStack supports what is called 'availability zones.' Availability zones help in supporting application higher availability and resilience. They ensure the physical independence of redundant application resources when assigned to different availability zones. Physical independence could mean separate power supply, network equipment, physical location, etc.

OpenStack has implemented the remote attestation principle of TCG specifications, as discussed in Section 6.2. The implementation of the remote attestation principle allows users to specify the trust level of servers when hosting their resources.

12.1.4 OpenStack Configuration Files

OpenStack configurations are managed via text-based files. The files are stored in the following directory: /etc/SERVICE_NAME/SERVICE_NAME.conf, where SERVICE_NAME is a variable representing the name of an OpenStack service. For example, the nova configuration file is stored by default in /etc/nova/nova.conf and it has the following type of configuration parameters: # LOGS, # AUTHENTICATION, # SCHEDULER, # VOLUMES, # DATABASE, # COMPUTE, # APIS, # GLANCE, and # NETWORK. OpenStack services run by a set of daemons that have a name starting with the service name, such as nova-*, glance-*, keystone-*, etc. They could either run on a single machine or be spread across multiple machines.

12.2 Amazon Web Services

Amazon Web Services (AWS) is a Cloud management platform that provides computing resources and services on a pay-per-use model. Amazon AWS and OpenStack have some comparable features. We could say that OpenStack provides almost all services that AWS provides and, in addition, OpenStack provides additional features that AWS cannot cover. This is because AWS targets a production environment, that is it can only support stable and proven-to-work technology that has been tested to work reliably in critical infrastructures.

OpenStack, in contrast, is an open-source research-oriented project aiming to establish the next generation of trustworthy Cloud computing. Unlike AWS, OpenStack as a result is still considered not production-ready. We now list the most important services in AWS and how they map to those of OpenStack.

Amazon Elastic Compute Cloud (Amazon EC2) provides computing resources in AWS. It is somewhat similar to Nova compute in OpenStack. AWS uses different naming conventions from those of OpenStack; for example, a VMI in AWS is called an Amazon machine image (AMI) and a VM instance is called an Amazon EC2 instance.

Amazon supports three types of storage systems, as illustrated in Figure 12.5: Amazon Elastic Block Store (Amazon EBS), Amazon Simple Storage Service (Amazon S3), and Amazon EC2 instance store. All these storage systems are available at OpenStack. Amazon EBS is persistent storage, which is similar to Nova-volume in OpenStack. Such volumes are not affected by an instance lifecycle. That is, if the instance is terminated for any reason, then its attached Amazon EBS volumes keep the data intact. OpenStack associates with each instance a local storage which is deleted when its instance is terminated. This is similar to the Amazon EC2 instance store. Amazon S3 is similar to the OpenStack Swift component. It provides access to a reliable and inexpensive data storage infrastructure.

The security measures provided by Amazon are also covered by OpenStack. OpenStack Keystone is called Identity and Access Management (IAM) by AWS. IAM manages users, federated users, and roles. AWS is also similar to OpenStack when controlling access to

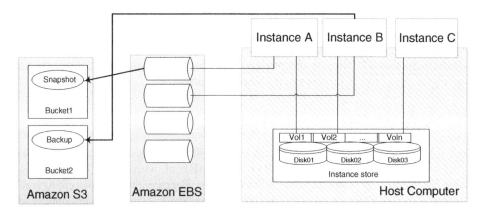

Figure 12.5 Amazon storage systems

network traffic. Users define appropriate firewall rules controlling network traffic to their VMs. In addition, AWS as in OpenStack supports availability zones to split redundant resources across separate physical resources. As indicated earlier, OpenStack provides richer features than those provided in AWS. For example, unlike OpenStack, AWS does not yet support the assessment of servers' trustworthiness.

12.3   Component Architecture

Figure 12.6 presents a high-level architecture which illustrates the main entities and a general layout of a scheme framework that implements some of the concepts discussed throughout the book. We use OpenStack controller node (i.e., the VCC) and OpenStack nova-compute (i.e., a computing node at the physical layer). The computing node runs a hypervisor which manages a set of VMs. The VCC receives two main inputs: user requirements and infrastructure properties. The VCC manages user virtual resources based on such an input. This section introduces new components to OpenStack. Adding a new component to OpenStack requires updating the following components: *nova-api*, *nova-database*, *nova-scheduler*, and *nova-compute*. In this section we present the modifications that are introduced on these components.

12.3.1   Nova-api

Nova-api is a set of command lines and graphical interfaces which are used by Cloud customers when managing their resources at the Cloud, and are also used by Cloud administrators when managing the Cloud virtual infrastructure. We updated *nova-api* library to consider the following:

- *Infrastructure properties.* The Cloud physical infrastructure is very well organized and managed, and its organization and management associate its components with infrastructure properties. Examples of such properties include: a resource chain of trust, components

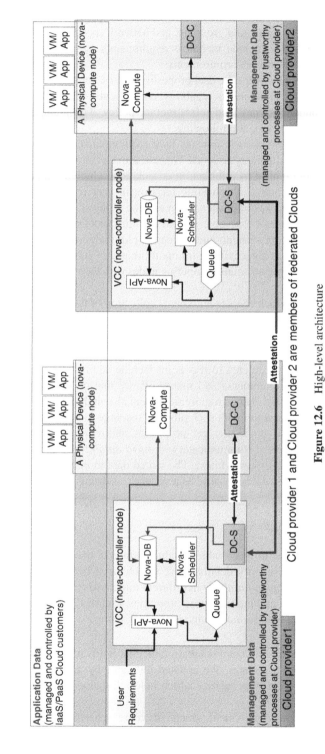

Figure 12.6 High-level architecture

reliability and connectivity, components distribution across Cloud infrastructure, redundancy types, servers clustering and grouping, and network speed.
- *User requirements.* These include technical requirements, service level agreement, and user-centric security and privacy requirements.
- *Changes.* These represent changes in user properties (e.g., security/privacy settings), infrastructure properties (e.g., components reliability, components distribution across the infrastructure, and redundancy type), and infrastructure policy. The main changes which we introduced at *nova-api* include the following:
  - Add an option to enable users to manage their requirements, which include but are not limited to security and privacy aspects. We tested our prototype on the following user requirements: geographical location and user isolation, which are managed manually at this stage. The first controls the hosting location of user resources, while the second controls the exclusion of certain users from sharing a physical server. It is important to stress that (at this stage) we assume system administrators are trusted. Planned future work will provide a more stringent mechanism to eliminate such an assumption.
  - Add an option which enables administrators to manage Cloud infrastructure properties and policies, for example associate computing nodes with their domains and collaborating domains.
  - Provide an interface which enables automated collection of the properties of the physical resources through trustworthy channels – at this stage we focus specifically on automating the collection of RCoT.

The main changes which are related to this point include adding an option to enable users to add a wide range of requirements and manage them; that is, updating nova-manage to enable users to create, list, and change requirements, and also updating the same library (nova-manage) to enable the association with each physical resource of a set of properties using the command lines options add_properties, get_properties, and remove_properties as illustrated in Figure 12.7. These libraries communicate with *nova-database* to store user choices and infrastructure properties, as discussed next. The stored data in *nova-database* is used by the presented scheduler ACaaS (access control as a service), as discussed next.

12.3.2   Nova-database

Nova-database is composed of many tables holding details of the Cloud components. It also holds users, projects, and security details (these get changed between different releases of OpenStack). We extended *nova-database* in different directions to maintain the taxonomy of Clouds, user requirements, and infrastructure properties. Figure 12.8 illustrates the modifications at *nova-database* in bold format, which are as follows:

- *Compute_nodes* is an existing nova-database table that holds records reflecting a computing resource at the physical layer. We updated this table by adding the following fields: RCoT(Physical) and security properties which hold a list of computing resource security details.
- *Requirement_entries and requirement_specification* are two new tables. The first holds the expected type of requirement a user is allowed to enter, such as federated Cloud, location,

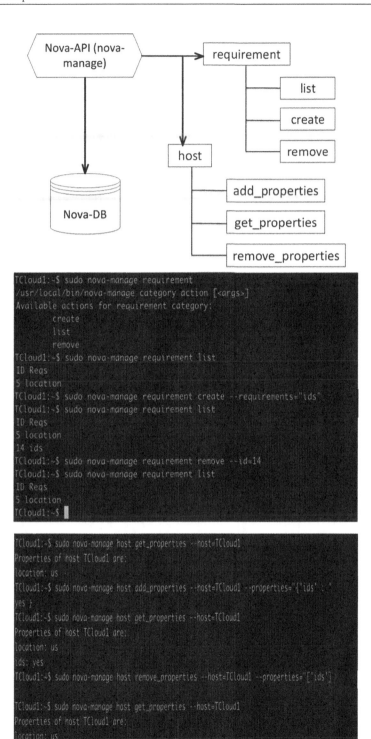

Figure 12.7   Nova-api updates

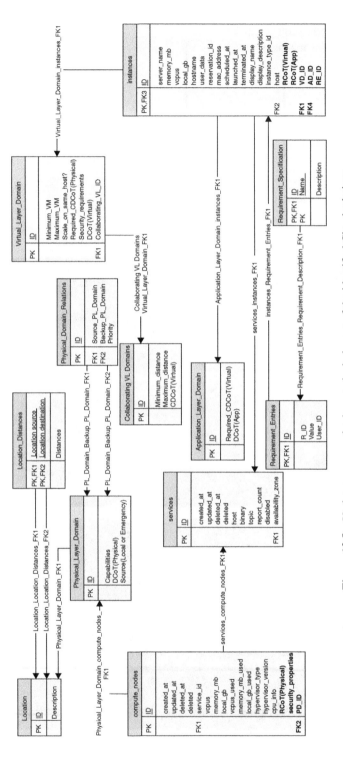

Figure 12.8 Updates on nova-database to include part of the identified Cloud relations

excluded user list, etc. The second holds a user-specific value of the defined requirements. Figure 12.7 demonstrates user interactions with these tables via the nova-manage command line interface.

- *Physical_Layer_Domain* is a new table which holds the records of Cloud physical domains. This covers both the primary Cloud provider physical domains and all other physical member domains of the emergency domain. The table defines the relationship amongst resources and holds physical domain metadata. The metadata includes the domain capabilities, DCoT, and a foreign key pointing to the table which identifies the relative geographical location of the physical domain within the Clouds and federated Clouds infrastructure. The table has a field source, which can have two values: local or emergency. Emergency means it is a federated Clouds domain while local means it belongs to the same Cloud infrastructure.
- *Location and Location_Distances*. The aim of these tables is to identify all possible *locations* at the Cloud infrastructure. They also define the relative distance between pairs of all identified locations. These tables are bound as follows: the compute_node table is bound to the physical_layer_domain table and the physical_layer_domain table is bound to a specific location identifier in the *location* table. The latter is bound to the location_distances table, which specifies all distances between a location identifier and all other location identifiers. In this we assume the resources of a physical domain are within close physical proximity, which reflects current deployment scenarios in practice.
- *Collaborating_PL_Domain* is a new table which establishes the concept of collaborating physical domains. Each record identifies a specific backup domain for each physical domain with a priority value. A source domain can have many backup domains. The value of the priority field identifies the order in which physical backup domains could possibly be allocated to serve as source domain needs. Backup domains are used in maintenance windows, emergencies, load balancing, etc. Backup domains should have the same capabilities and DCoT as the source physical domain itself.
- *Instances* is an existing OpenStack table representing the running instances at computing nodes. We updated the table by adding the following fields: virtual resource chain of trust RCoT(Virtual); application resource chain of trust RCoT(Application); two foreign keys which establish a relationship with the instance's virtual and application domain tables, as defined in the *Virtual_Layer_Domain* and *Application_Layer_Domain* tables, respectively; and RS_ID which is a foreign key pointing to the requirement_entries table.
- *Services table* is an existing OpenStack table which binds the virtual layer resources to their hosting resources at the physical layer.
- *Other tables*. Openstack has many more tables, which are beyond the scope of this chapter to discuss.

Most of the *nova-database* records are uploaded automatically, using the software agents as discussed in Chapter 9, the modified *nova-api*, and/or via OpenStack management tools. Ideally, such records should be securely collected and managed. At this stage our focus is on providing high-level architecture design, providing a running Cloud scheduler, and providing software agents that can attest to the trustworthiness of OpenStack components and then push the result to the nova-database. Full automation of Cloud management services is our planned long-term objective.

12.3.3   Nova-scheduler

In OpenStack, *nova-scheduler* controls the hosting of VMs at physical resources considering user requirements and infrastructure properties. Current implementations of *nova-scheduler* do not consider the entire Cloud infrastructure, nor do they consider the overall user and infrastructure properties. According to OpenStack documentation, nova-scheduler is still immature and great efforts are still required to improve it. We implement a new scheduler algorithm, ACaaS, which performs the following when allocating physical resources to host virtual resources: considers the discussed Cloud taxonomy; selects a physical domain's resource which has physical infrastructure properties that can best match user properties; and ensures that the user requirements are continually maintained. ACaaS collaborates with the following software agents (see Figure 12.6):

- *Cloud client agent, DC-C.* Runs at OpenStack computing nodes and performs the following: calculates the computing node RCoT and continually assesses the status of the computing node and passes the result over to DC-S; manages domains and collaborating member domains based on policies distributed by DC-S (e.g., a VM can only operate with a known value of a chain of trust and when the hosting physical collaborating domains have a specific value of CDCoT(Physical) as defined by user properties).
- *Cloud server agent, DC-S.* Runs at OpenStack domain controller and performs the following: maintains and manages OpenStack components (including the nova-scheduler) by ensuring they operate the Cloud only when they are trusted to behave as expected; manages the membership of the physical and virtual domains; and attests to DC-C trustworthiness when its computing node joins a physical domain. DC-S also intermediates the communication between DC-C and nova-scheduler, attests to DC-C's computing-node trustworthiness, collects the computing node RCoT, and then calculates DCoT, CDCoT, and stores the result in an appropriate field in nova-database.

12.4   Prototype

Having defined a high-level architecture of the scheme, this section describes a possible prototyping. We present a mechanism for a trustworthy collection of resource chains of trust, and then calculate for each group of resources their domain and collaborating domain chain of trust. Subsequently, we use the ACaaS scheduler to match user properties with infrastructure properties. Other infrastructure properties are either collected automatically (such as the capabilities of physical resources) or entered manually (such as the physical location of computing nodes). These properties could be altered by system administrators. The trust measurements performed by the DC-C identify the building up of a resource's chains of trust and its integrity measurements. This section discusses the implementation of the scheme framework. Our implementation also includes trust establishment building on remote attestation and secure scheduling.

12.4.1   Trust Attestation via the DC-C

The implementation is based on an open-source trusted computing infrastructure which is built on a Linux operating system. Building the resource chain of trust of a computing node

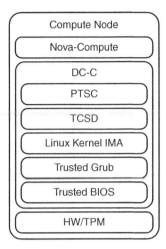

Figure 12.9   Compute node architecture

starts from the node TPM and ends with the node DC-C, as illustrated in Figure 12.9. The RCoT building process starts with the platform bootstrapping procedure, which initializes the TPM via a trusted BIOS. The trusted BIOS measures and loads the trusted bootloader [6] (i.e., the Trusted Grub), which measures and loads a Linux kernel. We updated the Linux kernel ensuring that the IBM Integrity Measurement Architecture (IMA) [7] is enabled by default. The IMA measures all critical components before loading them. These include kernel modules, user applications, and associated configuration files. The values of these measurements are irreversibly stored inside the PCRs of the computing node, which are protected by the TPM. The IMA by default uses PCR #10 to store the measurements.

The TPM driver and the Trusted Core Service Daemon (TCSD) [8] expose the Trusted Computing Services (TCS) to applications. These components constitute the part of the DC-C for collecting and reporting the trust measurement of a resource. The resource chain of trust is, hence, constructed from the CRTM [9], which itself resides in and is protected by the trusted BIOS.

Table 12.1 illustrates part of the records of the bootstrapping process for the prototype as generated by the IMA measurement log. The IMA measurement log is the source for generating integrity reports (IRs), which are used, as we discuss later, to determine the genuine properties of a target system during the remote attestation process. The first column in Table 12.1 shows the value of $PCR10$ after loading the components of the third column. The second column records the hash value of the loaded component. The first record holds the value of *boot_aggregate*, which is a combined hash value of $PCR0-PCR7$; that is, it possesses the measurement of the trusted computing base (TCB) of a computing node, including the trusted BIOS, the trusted bootloader, and the image of the Linux kernel together with its initial ram-disk and kernel arguments. Whenever a software component is loaded, the IMA module generates a hash value of the loaded component and then *extends* it into $PCR10$ by invoking the *TPM_Extend* command [10]. Such a command updates $PCR10$ to reflect the loaded component as follows: $pVal_i = hash(pVal_{i-1}, hVal_i)$. Subsequent rows in the table present the measurement logs of the bootstrapping workflow at the adopted operating system,

Table 12.1 Compute node bootstrapping
measurement log

$PCR10$	HASH	Loaded component
$pVal_0$	$hVal_0$	boot_aggregate
$pVal_1$	$hVal_1$	/init
$pVal_2$	$hVal_2$	ld-linux-x86-64.so.2
$pVal_3$	$hVal_3$	libc.so.6
...	...	...
$pVal_i$	$hVal_i$	nova-compute.conf
...	...	...
$pVal_j$	$hVal_j$	python
...	...	...
$pVal_k$	$hVal_k$	nova-compute
$pVal_{k+1}$	$hVal_{k+1}$	libssl.so.1.0.0
...	...	...
$pVal_l$	$hVal_l$	nova.conf
...	...	...

Ubuntu 11.04. Other OpenStack components are then measured, which include the *nova-compute.conf* script, the *python* executable, the *nova-compute* executable, supporting libraries, and critical configuration files.

To reduce the complexity and focus on a practical Cloud deployment case, the prototype turns off all unnecessary services at the base system. As a result, the value of $PCR10$ does not change except if a new software module (e.g., a user program, kernel modules, or shared libraries) is loaded on the computing node. The loading process could be either good (e.g., a security patch) or malicious. In this case, the loaded software module would be measured and added to the log records. Such a measurement changes the value of $PCR10$.

Our prototype intentionally filters out the IMA measurements of VMs, that is the QEMU program in the prototype. This is because a VM CoT should be built on a compositional CoT; that is, the IMA measurements of a VM should not be considered as part of the TCB of a computing node. The measurements of a VM should rather be controlled by the IaaS Cloud user and not the Cloud provider, as this will likely raise the user's privacy concerns. This measurement process, in addition, would significantly increase the complexity of the trust management. If an exploited VM runs on a computing node, for example, to perform a malicious behavior on other components and applications, the properties of the computing node would change once the exploited VM started to affect the TCB components. In such a case, the DC-C will leave the physical domain; that is, the DC-C will stop operating and the VMs which are hosted at the infected computing node will be forced to migrate to another healthy computing node member of the same physical domain.

Finally, the DC-C collects the integrity measurement logs as recorded by the IMA, and generates an IR following the specifications of the platform trust service (PTS) interface [11]. The DC-C, as we discuss in the next subsection, sends the IR and the signed PCR values to the DC-S on request. In the prototype, this component is implemented by integrating the PTSC module from the OpenPTS [12].

12.4.2   Trust Management by the DC-S

This section starts by summarizing the high-level steps of the implemented part of the system workflow.[4] It then presents the prototyping details which are related to the DC-S. These are as follows:

- Cloud security administrators could either create a new physical domain or use an existing domain. The creation process involves deciding on the domain capabilities, location, and defining its collaborating domains. As discussed in Section 12.3, we updated *nova-api* to enable administrators to manage this process.
- Cloud security administrators then install the DC-C and nova-compute at all new physical computing nodes that are planned to join the domain.
- The DC-C joins the cloud physical domain by communicating with the DC-S. The DC-S would first attest to the DC-C's trustworthiness and establish an offline chain of trust with the DC-C (using sealing and remote attestation concepts, as proposed by TCG specifications). Next, the DC-C would calculate its host chain of trust RCoT, as described in Chapter 9, and pass the results to the DC-S.
- Subsequently, the DC-S would store the RCoT at the *compute_nodes* table, and ensure that all devices in each domain have the same capabilities. The sealing mechanism, which is established in previous steps, assures the DC-S that the DC-C can only operate with the same value of the reported RCoT. If this value changes (e.g., as in the case of the hosting device being hacked), the DC-C will not operate. This prevents VMs from starting at a hacked device.
- Users, using *nova-api* commands, deploy their VMs and associate them with certain properties. Such properties include, for example, the required CDCoT(Physical) and the multi-tenancy restrictions which control the sharing of a computing node with other users.
- The ACaaS scheduler allocates an appropriate physical domain to host a user VM. The properties of the physical domain and its member devices should satisfy the defined user requirements.

The remaining part of this section covers the implementation of the remote attestation process and the secure scheduling.

Remote Attestation

Our prototype implements the remote attestation process using OpenPTS [12] which is managed by the DC-S. OpenPTS sends an attestation request to each computing node to retrieve its IR and PCR values. When a computing node sends the requested values, OpenPTS would then examine the consistency of the received IR and PCR values [9]. Subsequently, it would verify the security properties of the computing node by matching the reported IR with the expected measurement from a white-list database [9]. The white-list database stores sets of measurements, where each set is calculated based on a carefully selected *good* platform

---

[4] Further details about these steps are provided in [13].

configuration state. The calculation is performed on a *good* platform in the form of hash values for a selected set of pre-loaded software components.

For the purpose of the prototype, we used two newly installed Ubuntu 11.04 servers. These servers have minimal settings for a computing node to perform its planned functions. The hash values of the software stack of each computing node should exist within the white-list database. If it does not, we consider the computing node to be untrusted. The *good* configurations could be extended/changed by adding/updating their corresponding values in the white-list database.

The attestation protocol works as follows. Every computing node (C_i) is identified by its AIK. The AIK is certified by the Cloud controller VCC (M) as it covers the Privacy-CA role [9, 14]. When a new computing node is added to the Cloud infrastructure it must first be registered at the VCC which then certifies its AIK. Only registered computing nodes can connect to the VCC as their certified AIKs cannot be forged and AIKs can only be used inside the genuine TPM that generates them. The registration steps of C_i at M are outlined in Protocol 12.1. Whenever a computing node sends a request to connect to the VCC a trust establishment protocol is executed which is outlined in Protocol 12.2.

Protocol 12.1 Computing Node Registration Protocol

A computing node (C_i) sends a registration request to VCC (M) as follows. First, C_i sends a request to its TPM to create an AIK key pair using the command TPM_CreateAIK. The TPM would then generate an AIK key pair. The generated private part of the key pair never leaves the TPM, and the corresponding public part of the key pair is signed by the TPM endorsement key (EK) [9]. The EK is protected by the TPM, and never leaves it. C_i then sends a registration request to M. The request is associated with the EK certificate, the AIK public key, and other parameters:

$$C_i \rightarrow M : Cert(K_{EK_i}), \{K_{AIK_i}\}_{K^{-1}_{EK_i}} \tag{12.1}$$

M certifies AIK_i as follows. M verifies $Cert(EK_i)$. If the verification succeeds, M generates a specific-AIK certificate for C_i and a unique ID, CID_i. It then sends the result to C_i:

$$M \rightarrow C_i : \{Cert(K_{AIK_i}), CID_i\}_{K^{-1}_M}, Cert(K_M) \tag{12.2}$$

Protocol 12.2 Trust Establishment Protocol

M sends an attestation request to C_i. The request includes a nonce N_a. C_i would then report an attestation ticket to M as follows. C_i sends its PCR values, and the measurement log IR back to M, together with N_a. These are signed using the C_i's AIK:

$$C_i \rightarrow M : \{N_a, \{PCR\}, IR\}_{K^{-1}_{AIK_i}} \tag{12.3}$$

M then verifies the message sent by C_i as follows. It verifies the AIK_i signature and N_a matches the sent nonce. If the verification succeeds, M examines the consistency of PCR and IR, and then determines the properties of C_i based on the value of IR.

The configurations of a computing node could possibly be altered after an attestation session, for example loading a new application. In such a case, the computing node attestation properties (as maintained by the VCC) would be violated. Addressing this would require establishing a trusted channel [15] to *seal* [9] the communication key with the verified PCR values. The sealing process provides the assurance that the DC-S can load the key only for a specific computing node's configuration. Any changes in the computing node configurations would trigger a new attestation request from the VCC to the computing node.

The implementation of the trusted channel, when sealed keys are loaded into memory, requires a small TCB. The TCB should enforce strict access to the memory area which stores the key. Having a large TCB, however, could result in leaking the key from memory without reflecting that on the platform trust status. Implementing a small TCB is a challenging problem, especially considering the complexity and scalability of the hosting Cloud system (we leave this important subject as planned future research). As an attempt to lessen the impact of this threat, in the prototype we impose periodic attestations which keep the security properties of a computing node up to date. We implemented this by associating a timer with each computing node. Re-attestation is enforced whenever the timer expires. Untrusted computing nodes found by the re-attestation will be removed immediately from the database and would need to re-enroll in the system for future use. In addition, VMs running on untrusted computing nodes will be forced to migrate to other computing nodes which are members of the same physical domain.

Secure Scheduling

As we discussed in previous sections, computing nodes are organized into physical domains. Such organization is based on the properties of each computing node (i.e., security, privacy and other properties) which enable it to serve the needs of the domain. Users can specify their expected properties of computing nodes that could host their VMs. Some of the properties could be represented by a set of PCR values. However, PCR values are hard to pre-calculate and manage as they represent aggregated hash values of software components when loaded in a specific order. In the prototype, users do not need to specify PCR values, rather they would need to identify their desired hosting environment using the provided sets of white-lists. A computing node white-list is identified in accordance with its properties which get attested whilst joining a domain and periodically thereafter. Genuine updates on the properties of a computing node (e.g., applying a security patch) require adjustment of the corresponding record in the white-list database. In the prototype, part of the user's required properties could represent entries in the white-list database. The ACaaS scheduler deploys each VM on a computing node that has the same properties as the one requested by the user of the VM. The ACaaS scheduler, in collaboration with the DC-S and DC-C, periodically examines the consistency of such properties.

12.5   Summary

Establishing the next generation of trustworthy Cloud infrastructure is a complex mission which requires collaborative efforts between industry and academia. This book presents the foundation of Cloud computing science building on solid in-depth and diverse experience in this domain. Part Two presented a set of integrated frameworks which form the roadmap for establishing trust in Clouds. It also presented a list of framework requirements and discussed possible solutions to some of the requirements. This chapter has presented a possible implementation of some components of such frameworks. We also introduced commercial and open-source Cloud management platforms: Amazon AWS and OpenStack.

References

[1] OpenSource. OpenStack, 2010. http://www.openstack.org/.

[2] RackSpace, 2013. http://www.rackspace.com.

[3] NASA, 2013. http://www.nasa.gov.

[4] R. Sandhu, E. Coyne, H. Feinstein, and C. Youman. Role-based access control modles. In *IEEE Computer*, pp. 38–47. IEEE, 1996.

[5] Sun Microsystems Inc. NFS: Network File System Protocol specification. RFC 1094, Internet Engineering Task Force, March 1989.

[6] Trusted Grub. http://trousers.sourceforge.net/grub.html.

[7] Reiner Sailer, Xiaolan Zhang, Trent Jaeger, and Leendert van Doorn. Design and implementation of a TCG-based integrity measurement architecture. In *Proceedings of the 13th Conference on USENIX Security Symposium – Volume 13, SSYM'04*, pp. 16–16. USENIX Association: Berkeley, CA, 2004.

[8] Trousers – the open-source TCG software stack. http://trousers.sourceforge.net/.

[9] Trusted Computing Group. http://www.trustedcomputinggroup.org.

[10] Trusted Computing Group. *TPM Main, Part 3, Commands. Specification version 1.2 Revision 103*, 2007.

[11] Infrastructure Work Group Platform Trust Services Interface specification, version 1.0. http://www.trustedcomputinggroup.org/resources/infrastructure_work_group_platform_trust_services_interface_specification_version_10, 2006.

[12] Open Platform Trusted Service User's Guide. http://iij.dl.sourceforge.jp/openpts/51879/userguide-0.2.4.pdf, 2011.

[13] Imad M. Abbadi. Clouds infrastructure taxonomy, properties, and management services. In Ajith Abraham, Jaime Lloret Mauri, John F. Buford, Junichi Suzuki, and Sabu M. Thampi (eds), *Advances in Computing and Communications*, vol. 193 of *Communications in Computer and Information Science*, pp. 406–420. Springer-Verlag: Berlin, 2011.

[14] Privacy ca. http://www.privacyca.com.

[15] Yacine Gasmi, Ahmad-Reza Sadeghi, Patrick Stewin, Martin Unger, and N. Asokan. Beyond secure channels. In *Proceedings of the 2007 ACM Workshop on Scalable Trusted Computing, STC '07*, pp. 30–40. ACM: New York, 2007.

13

Case Study

This chapter presents a case study of some of the concepts discussed in the book to address real-life challenges. The first section describes the scenario and subsequent sections address issues building on the scenario.

13.1 Scenario

This section describes a scenario for using a home healthcare system in Cloud computing. The scenario is based on a hospital which provides home services to clients. The services are accessible through web portals that are provided through the hospital's website. These services are hosted on a Cloud infrastructure using the IaaS service type. Users should not need to be aware of the existence of the Cloud, as all technicalities must be transparent to them. Users might include patients, care givers (patient family members), hospital staff (e.g., general practitioners, medical consultants, psychiatrists), and other collaborating organizations with the hospital (e.g., research centers).

System administrators of the Cloud allocate virtual resources and manage them based on a pre-agreed SLA with the hospital. This includes allocating VMs, virtual storage, networking, and managing them. The hospital, however, is in charge of installing and maintaining the operating system and all software packages which are needed to run the hospital application. For example, the hospital is in charge of maintaining the operating system, database management system, application servers, and developing and deploying the hospital application. The hospital can outsource this service to a professional IT services company, or can have its own IT staff to maintain the infrastructure provided by the Cloud provider. Once the hospital application is deployed on the Cloud, the hospital services are then made available to clients through a web page. The process should be completely transparent to clients; that is, clients should access the application by connecting to a URL provided by the hospital regardless of the existence of the Cloud. Clients will then use the credentials provided by the hospital to log in and access the allocated services.

The Cloud service provider can have SLAs with other third-party service providers (e.g., hardware suppliers and operating system vendors) to act as an escalation point for critical failures and to provide additional support for services that are not in house.

Cloud Management and Security, First Edition. Imad M. Abbadi.
© 2014 John Wiley & Sons, Ltd. Published 2014 by John Wiley & Sons, Ltd.
Companion Website: www.wiley.com/go/abbadi_cloud

13.2   Home Healthcare Architecture in the Cloud

In this section we present a possible architecture of the home healthcare system when deployed at a Cloud provider. This is illustrated in Figure 13.1, which also covers the required self-managed services. The architecture only covers the virtual and the application layers. In the application layer we have two types of domains as follows:

- *Application-specific domains.* The figure illustrates two domains specifically created for the hospital application and other dependent applications, such as the research center collaborating with the hospital. The number of domains could be more than that, based on several factors which we do not discuss here for simplicity.
- *Application services domains.* These are provided as PaaS. Examples could include the following: encryption service, log management service, and database management service.

In the virtual layer we created two domains: one to host the middle-tier component and the other for the backend component. Many other domains could be created. The properties of the domains are not discussed for simplicity.

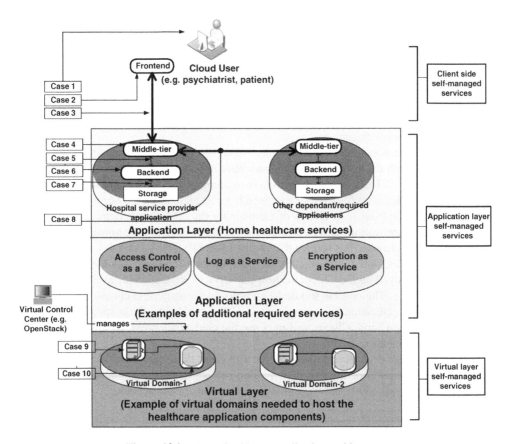

Figure 13.1   Home healthcare application architecture

13.3 Insiders Analysis for Home Healthcare

13.3.1 Model Instance

Based on the scenario description above, we can instantiate the model as shown in Figure 13.2. The *hospital* is an organization that needs Cloud computing resources and therefore will be the *Cloud customer*. A hospital has a number of employees, including: a *researcher, hospital system administrator*, and a *psychiatrist* who *work_on PatientRecord* (a type of asset for the hospital).

The hospital provides healthcare services to its patients, (i.e., clients), who co-own the patient record and are *cared_for* (a sub-relation of *Client related_to Client* in Figure 11.3) by their care givers. Hospitals may co-exist in the Cloud with other organizations (which we label as 'competitor') that may be interested in the hospital's assets.

13.3.2 Identifying Potential Insiders and Insiders

In this subsection we identify potential insiders and insiders using the process outlined in Section 11.3.4. The process starts from the identified system components given in Section 11.3.3; that is, a physical device, a hypervisor, a VM, and an application. In this section we identify potential insiders and insiders in each of these components, as follows.

Physical Device

A physical device can belong to three layers: storage, network, or server, as illustrated in Figure 13.3.

- *The storage layer*. At this layer the type of physical device would be a storage device, which is vulnerable to different types of threat. For example: it can be swapped with a corrupt device; taken away and mounted in another system; and the device's content can be copied or altered.

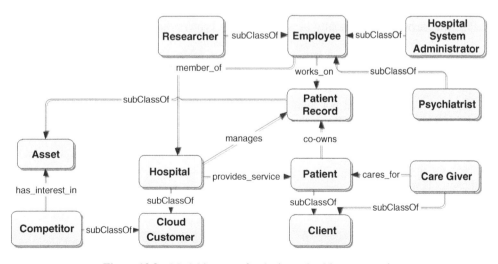

Figure 13.2 Model instance for the home healthcare scenario

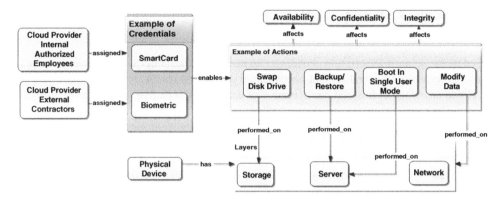

Figure 13.3 Physical infrastructure

Based on the attack scenario, these threats could have an impact on content confidentiality, integrity, and/or availability. For example, content may lose integrity through backup/restore operations, content may lose availability through removing a device's content, and content could be leaked by copying it to a USB memory stick.

- *The server layer.* At this layer the physical device type would be a physical server, which is vulnerable to different types of threat. For example: the physical server is vulnerable to all possible hardware threats, and it can be started in a different configuration from that expected, for example by booting it in single-user mode. This could affect content availability, integrity, and confidentiality based on the attack scenario; for example, booting the server in a single-user mode enables attackers to access the superuser account without the need to possess any authorization credentials.
- *The network layer.* At this layer the physical device type would be a network component. Data can be modified as it is transmitted to/from the device, affecting availability, confidentiality, as well as integrity.

Physical devices will normally be stored in data centers that have access restrictions to few individuals. This access is typically enforced using credentials such as biometrics and smartcards. Such credentials, will be assigned to authorized employees from Cloud service providers (e.g., system administrators) and employees from organizations contracted by the Cloud service provider.

Based on the insider and potential insider definitions, Cloud authorized employees and contractors are potential insiders as they are provided with credentials that can access physical devices. Once the potential insiders use the credentials and cause harm, then they are insiders. Also, anyone who has access to these credentials (by stealing them, or the system administrator himself sharing them with unauthorized persons) is considered an insider once he uses them and causes harm.

Hypervisor

The hypervisor runs a VMM which controls the VMs running on the physical device. It comprises a thin-layer kernel and management services, as illustrated in Figure 13.4. The

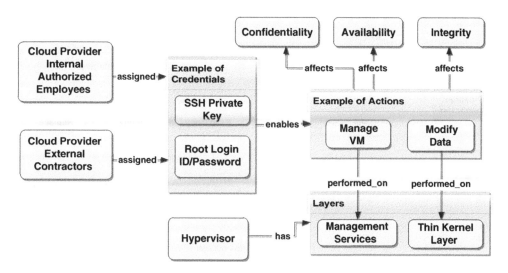

Figure 13.4   Hypervisor component breakdown

management services enable the VMs' management actions, such as start, stop, and migrate, to be performed. Because network traffic to and from the VMs is mediated by the hypervisor, data for the VMs can also be modified through the hypervisor. All these actions may impact the availability, integrity, and confidentiality of the services offered by the hospital. The typical credentials that enable accessing the hypervisor to perform such actions include root login credentials and SSH private keys. *Cloud provider authorized employees (e.g., system administrators) and contractors* are the main actors that are expected to be assigned these credentials.

Based on the insider and potential insider definitions, Cloud authorized employees and contractors are potential insiders as they are provided with credentials that can access the hypervisor. Once the potential insiders use the credentials and cause harm, they are insiders. Also, anyone who has access to these credentials (by stealing them, or the system administrator himself sharing them with unauthorized persons) is considered an insider once he uses them and causes harm.

Virtual Machine

VMs are containers that comprise an operating system and applications. These are stored together with configuration information in a disk image. Figure 13.5 shows examples of actions that may be performed on any of the layers, which could affect all three security properties (i.e., availability, confidentiality, and integrity). For example, updating binaries can be performed on the operating system and application, which can affect the three security properties. The entire disk can also be copied, affecting the confidentiality of the stored data. The data might be modified, affecting its availability and integrity.

Two types of credential would typically be needed to perform the identified actions: the SSH private key and the root login id/password. These would enable all the actions identified

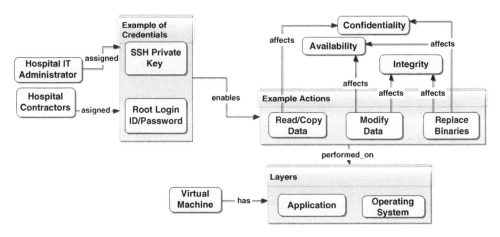

Figure 13.5   VM access

at all layers. *Hospital internal system administrators and contractors* working on behalf of the hospital would be the main actors expected to be assigned root login and SSH private keys. However, system administrators from Cloud service providers of IaaS type should not normally get root access to the VMs.

Based on the insider definition in Chapter 11, hospital Cloud internal system administrators and contractors could be potential insiders as they are provided with credentials that can access the main patient information repository from a server-side application. Also, anyone who has access to a system administrator authorized authentication credential is considered a potential insider (by stealing it, or the system administrator himself sharing it with unauthorized persons).

Based on the insider and potential insider definitions in Chapter 11, hospital internal system administrators and contractors are potential insiders as they are provided with credentials that can access VMs. Once the potential insiders use the credentials and cause harm, then they are insiders. Also, anyone who has access to these credentials (by stealing them, or the system administrator himself sharing them with unauthorized persons) is considered an insider once he uses them and causes harm.

Application

Applications run on VMs and can be either client-side applications or server-side applications, as illustrated in Figure 13.6. These are stored and run on VMs. Figure 13.5 shows examples of actions that may be performed on any of the layers, which could affect all three security properties (i.e., availability, confidentiality, and integrity). For example, modifying data can be performed from client-side or server-side applications, and it can also affect the three security properties. Content stored in the server-side application can be copied (affecting data confidentiality), altered (affecting data integrity), or removed (affecting data availability).

End-users would be assigned user logins allowing them to perform actions enabled by this credential. Examples of such users include patients, care givers, and hospital employees (e.g., researchers and psychiatrists).

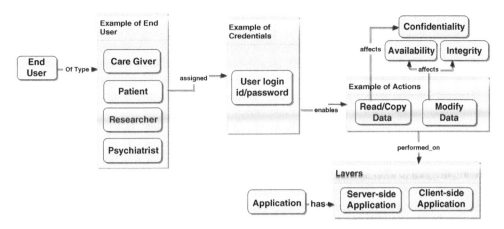

Figure 13.6 Application access

Based on the definition of insiders in Chapter 11, end-users could be potential insiders as they can access data using authorized credentials. Also, anyone who has access to an authorized authentication credential (by stealing it, or the authorized user himself sharing it with unauthorized persons) is considered a potential insider.

Based on the insider and potential insider definitions in Chapter 11, end-users are potential insiders as they are provided with credentials that can access the application's content. Once the potential insiders use the credentials and cause harm, then they are insiders. Also, anyone who has access to these credentials (by stealing them or the end-user himself sharing them with unauthorized persons) is considered an insider once he uses them and causes harm.

13.3.3   Insider Threat Analysis

Insiders' actions could affect information/service availability, integrity, and confidentiality. The proposed methods at the time of writing for addressing insider threats focus mainly on mitigating the threats to content confidentiality. In our opinion, the lack of schemes addressing insider threats to content integrity and availability is due to two main reasons: the lack of solid cases discussing insider threats and the nature of the problem, which is not easy to address as insiders must be authorized to update/remove records.

The possible threats that can be raised by the identified insiders in Section 13.3.2 are as follows:

- *End-users*. These access the hospital application services via a provided authentication credential. Each user is assigned a credential with access rights for accessing the provided services, which enables a user to create new records, update patient records, and delete patient records. Such rights should not provide the user the ability to access the system from the backend (i.e., from the operating system level or database management system level), and they should not provide users with the ability to have a global effect on services (e.g., stop a service or remove the whole data repository).

The insider threats of end-users are restricted to the granted access rights that are provided to the end-user credential. For example, if access rights allow the user to only read a patient record, then the insider threat affects content confidentiality. If access rights allow the user to update and delete a patient record, then the insider threat affects content integrity and availability, and so on.

- *Hospital internal system administrators.* These access the hospital application and back-end virtual resources via a provided authentication credential(s). A system administrator is in charge of maintaining the application and backend services (e.g., operating system and database management system). System administrator are assigned access rights for performing their job, which could enable them to perform critical actions on the system (e.g., suspend a VM, backup/restore operations, migrate VMs, and stop/restart middle-tier application servers). Such rights enable their holder to have a global effect on the provided hospital services (e.g., to stop a service, remove the whole data repository, or leak the data repository for patient records).

  Insider risks in this case would be based not only on the access rights that are provided to the account used by the insider but also on the security best practices used (e.g., separation of duty and least-privilege concepts). For example, an organization might reduce the impact of data integrity by introducing a database/application backup role, which is separate from the system administrator role. Also, an organization can introduce an application maintenance role that is separate from the database management role. The application of security best practice does not necessarily prevent insider threats, but it will lessen their effects. We now list the main insider threats for the system administrator role.

  – *Availability.* An insider can affect system availability. For example, the application management role can stop/delete middle-tier application services, the database management role can stop/delete the database, and the operating system role can stop the virtual resources. All these are examples of how an insider can cause a global effect on service availability.

  – *Integrity.* An insider can affect system integrity. For example, the application management role can create an authorized user account for a non-existent general practitioner, update patient records, and then delete the account. A backup role can invalidate the backup. A database management role can update patient records directly from the database.

  – *Confidentiality.* An insider can leak sensitive content to unauthorized parties. For example, a backup role grantee can copy the backup to a memory stick, restore it at home, and then leak the content to others. A database management role can also copy the database to a memory stick, or even search and then extract selected patient records to a USB stick or leak them via email.

- *Hospital contractors.* Hospital contractors are provided with appropriate credentials enabling them to maintain part of the hospital provided services (e.g., application support, operating system, and database management system). Contractors should be assigned the minimal access rights that are sufficient to do the job. Such rights could enable them to carry out critical actions on the system, exactly as those described for the system administrator role. Insider threats caused by external Cloud contractors have the same severity level as those caused by internal system administrators. Identifying these would be based on the roles granted to the contractor.

- *Cloud provider internal employees.* These have full access to the physical hardware resources (servers, storage, and network devices) and the operating system (hypervisor), which serve the provided virtual resources. In addition, they have full access to the Cloud infrastructure management software packages. These are used to maintain and monitor the virtual resources, for example stopping, starting, suspending, resuming, migrating and backing up a VM, and allocating/revoking computational resources to/from a VM.

The insider threats caused by the Cloud provider insiders could have a greater effect than the hospital insiders. This is because insiders could have even more authoritative access to the underlying infrastructure. Also, they are the parties who manage the hospital allocated virtual resources. In the following we briefly outline these threats:

- *Insider threats that affects content availability*. An insider who is granted a virtual resource management role can deprive some of the computational resources that are granted to the VM, which cause the machine to be non-responsive, for example, in peak periods.
- *Insider threats that affect content integrity*. An insider who is granted access to the hypervisor layer as a super-user can access the VM running on the hypervisor, enabling the insider to update the VM content. Also, an insider can restore VM storage from an old/hacked backup.
- *Insider threats that affect content confidentiality*. An insider with proper access privileges can copy a VM image or a backup from the storage server and restore these at home, which enables the insider to leak the hospital patient information to unauthorized parties.

- *Cloud provider external contractors*. Cloud provider external contractors are provided with appropriate credentials enabling them to maintain part of the Cloud infrastructure (e.g., hardware suppliers and software application support). Contractors should be assigned the minimal access rights that are sufficient to do the job. Such rights could enable them to perform critical actions on the system, exactly as those described for the Cloud internal employees. For example, a contractor that maintains the storage can perform backup of the storage and restore it at home, which enables him to leak sensitive content. Insider threats caused by external Cloud contractors have the same severity level as those caused by Cloud internal employees. Identifying these would be based on the roles granted to the contractor.

- *Cloud-of-Cloud internal employees*. As discussed before, if two Cloud providers collaborate, one Cloud's internal employees could access another Cloud's data that migrates across to their internal infrastructure. In this case the destination Cloud provider's system internal employees can cause the same level of threats 'on the migrated data' as the source Cloud provider internal employees, as discussed in the previous point.

- *Cloud provider customers*. Cloud provider customers in multi-tenant architecture [1] organizations share the same hardware resources. Here, all the employees of an organization who are authorized to access their organizational resources in the Cloud might be insiders for other organizations sharing the same hardware resources. For example, an attacker can learn sensitive information about other organizations (e.g., by exploiting covert channels [2, 3]).

13.4 Cloud Threats

In the previous section we discussed the effects of insiders on the home healthcare application when moving into the Cloud. We found that the insider threats when moving into the Cloud exceed those prior to moving to the Cloud. This section discusses additional threats that could face the home healthcare application when running in the Cloud. The following is a non-exhaustive list of potential issues:

- *Federated Cloud and third parties.* All the challenges faced by outsourcing the home healthcare application to the public Cloud also apply to federated Cloud partners with additional difficulties, as follows:
  - Where contractual agreements are legally binding with the primary Cloud provider, would the same contractual arrangements hold true with any federated partners and third parties?
  - How would contractual arrangements between Cloud partners affect the home healthcare system? For example, are legal contracts unenforceable within the UK? Are security controls between federated Cloud partners the same?
  - Would the home healthcare information assets be protected in the same fashion across the whole of the federated Cloud structure? If not, why not and how could this be contractually enforced? If so, how can assurance of this be gained?
  - What could prevent partnering Clouds establishing further federated services with third parties and using these new partners to store and process the home healthcare information assets in an unapproved manner? Furthermore, even if third parties were to be engaged within the partnering Cloud and not used to process the healthcare assets, the third parties would be considered an additional threat to the hospital by virtue of newly shared infrastructure and a potential attack vector arising from logical connection. How could the hospital be assured that the third parties are suitably separated from the hospital assets upon the approved public Cloud provider's infrastructure?
- *Physical security of Cloud provider sites.* It is likely that the home healthcare information assets will be held in data centers with access available from the public Cloud provider and their satellite administrative sites. It is therefore challenging to ensure that these parties apply and demonstrate physical security compliance in a manner commensurate with those within the hospital itself.
- *Personnel security.* It is unsure what personnel security checks may occur as part of the Cloud provider's employment procedures and furthermore, where pre-employment screening takes place, are the checks considered to be commensurate with the checks performed by the hospital?
- *The insider threat.* As described earlier, when outsourcing the home healthcare application to a public Cloud, the number of potential insiders increases from merely those considered within the hospital itself to an unknown proportion and with it the probability that a security breach may occur at their hands.
- *Architectural and technical security controls.* It is considered to be a significant challenge to ensure that a Cloud provider deploys the following technical security mechanisms, in a manner commensurate with the hospital, to protect the security of its patients. Principles of defense in depth should be employed to provide a layered security approach to detect, delay, and repulse a threat actor.
- *Identification and authentication.* Applying principles of defense in depth will require all hosts (including privileged users), application services, and data transfer partners to correctly authenticate each other. This is to prevent attacks that might hijack the service or capture data in transit. Methods for strong ID&A are made troublesome by password policies such as password duration, password complexity (will the technology support complex passwords?), password reuse, etc. Furthermore, a number of strong methods – including

two-factor authentication using cryptographic means such as digital certificates and smart tokens – bring with them key management issues to overcome.

- *Access control policies.* Access control policies would need to be in place to enforce the principle of least privilege and need to know. A challenge to be overcome is if it could be implemented such that only the hospital administrators and their clients had access to the home healthcare information assets residing within the Cloud by using an appropriate access control mechanism.

- *Server hardening.* All servers employed should be evaluated by common criteria to at least EAL4 and should be hardened in accordance with the common criteria security target and set to fail secure.

- *The network security.* The network infrastructure should adopt a layered model to ensure secure data separation boundaries between layers. Network hosts should be hardened according to their common criteria security target and set to fail secure. Additionally, a mix of disparate routers, switches, firewalls, etc. should be used. This is to ensure that if one is compromised, the same attack technique cannot be used to defeat all hosts in the architecture.

- *O/S hardening.* All O/S employed should be hardened as per suitable guidelines and configured for use in accordance with their common criteria security target.

- *Virtualization hardening.* Of particular importance is the hardening and configuration of hypervisors and individual VMs. Virtual machines are considered information assets in parallel with their physical counterparts. When incorrectly configured, the use of virtualization can weaken security controls in the guest O/S and fail to provide secure data separation. Even where a guest O/S has been hardened, an incorrectly configured hypervisor can allow privilege escalation. Virtualization products can be a single point of attack focus. The virtualization chosen must have undergone common criteria evaluation and be configured according to their security target. Furthermore, all virtualized machines and software must be patched in accordance with their non-virtualized counterparts to aid prevention of threats intended to exploit published vulnerabilities.

- *Data confidentiality.* Whilst all the controls listed refer to information specifically relating to software application asset security (C, I and A), one must be mindful that a public Cloud provider must show due diligence when storing hospital patient information. Should this information be misused, it could affect patient life.

- *Secure deletion of information.* When information is deleted, it must be permanently taken off the public Cloud provider's technology, rendering it computationally improbable that the data could be retrieved using popular forensic tools.

- *Protective monitoring.* Protective monitoring is vitally important to detect if a security breach has occurred and to positively identify the culprit such that they cannot deny doing the deed (non-repudiation). Protective monitoring is a primary feed into the incident management process and can also be used to ascertain correct service billing. All audit logs should be centrally held (but in split data stores for resilience), and held under strict access control utilizing principles of segregation of duty for authorization and for access to force collusion of two parties to affect a security breach. It is important, and something of a challenge, that the time is synchronized across the whole of the application technical infrastructure. Whilst this is technically possible (synch to a UTC time source), whether all the federated Cloud members would have joined technical policies to effect this is questionable. Furthermore, where the time is synchronized to a centralized source located in the USA, would this be considered acceptable evidentially during forensic investigations leading to litigation within the disparate jurisdictions (e.g., would the Chinese legal system accept a US time source to be acceptable corroborative evidence)? Additionally, sufficient events must be captured to ensure the logs are worthwhile and that the logs are made available to the hospital upon its request to cross-check access, billing and to aid forensic examination post-security incident. Additional uses for protective monitoring, when deployed in a comprehensive fashion, could be to generate metrics aiding measurement of the effectiveness of self-managed services (availability) in terms of MTTD, MTTI, and MTTR.

- *Procedural controls.* It is important to ensure that the Cloud provider is bound within procedural controls such as segregation of duty, separation of duty, security awareness, need to know, least privilege, and two-man rules (to name but a few). Such controls can ensure that security breaches are detected, and will often force collusion for a security breach to be effected. Such controls can be enforced by the deployment of individual 'Forms of Understanding' tied into contractual terms and conditions of employees and third parties within the supply chain.
- *Business continuity.*
  - *Backup and restore.* It is important that, should a security incident occur, the hospital services can be returned to normal within a period of time according to a predefined service level agreement. With this arises the challenge of backup storage, location of backup resources, and the necessary protection offered to these locations and resources.
  - *Security incident management and crisis management.* Where a security incident occurs to the home healthcare application or to the Cloud provider it is vitally important that incident management procedures are in place not only to effect resolution of the problem but also to ensure that correct lines of communication are issued between all parties with escalation to a recognized CERT (GovCERTUK). A problem for the Cloud provider may constitute a problem for the hospital and, potentially, vice-versa. How can this challenge be overcome? Perhaps agreement over communication and incident management operating procedures could be encapsulated into the security operating procedures which are subsumed into legally binding contracts.
  - *Resilience.* Resilience enhances the availability of the information security triad and ensures that the service is available when the business demands. A resilient design must be proportional to the criticality of the system.
  - *Elasticity.* As resource demands grow and shrink, ensuring that there is no waste in resources available (that all resources are utilized) and that in times of high demand, resources will automatically be assigned to prevent availability issues. The challenge is to ensure that resource provisioning is correctly ascertained initially and that billing is correctly calculated.
- *Non-comparable legal models/disparate jurisdictions.* A public Cloud provider may operate over a number of different countries; while some countries may share common legal aspects, some may not and may have their own individual requirements. The following is a non-exhaustive list of potential issues:
  - *Statutory compliance.* Data protection legislation may not exist in some countries in which the hospital healthcare assets are stored. Information could be stored by a public Cloud provider (or its third-party suppliers) within an area not party to legislation akin to the UK Data Protection Act or the US Safe Harbor Frameworks.
  - *Protection of intellectual property rights (IPR).* Protection of IPR and strict data ownership may not exist or may not be subject to sufficient enforcement within some countries in

which the public Cloud provider or an authorized third-party supplier processing the healthcare IPR operates (e.g., non-members of the World Trade Organisation).

– *Privacy laws*. Within some countries (such as Germany and Switzerland), privacy laws may be such that the implementation of detective measures – such as comprehensive protective monitoring policies – may breach the privacy of the users. Rigorous privacy laws make it difficult to deploy certain monitoring tools and to forensically prove an individual's actions leading to a security breach with non-repudiation.

– *Use of technology as preventative, detective and reactive measures*. Within some countries a combination of privacy laws and national security may prohibit some preventative and detective technical controls from being deployed. Privacy laws prohibit some accounting and audit technical controls (as mentioned within the previous point). Individual country's national security laws may prohibit or enforce the use of preventative technical controls such as cryptographic products. France limits the importation of certain cryptographic products while the USA, used to consider strong cryptographic products as munitions.

• *Unknown threat landscape development*. Where home healthcare is outsourced to a public Cloud, one can assume that the threat landscape will change and due to the global nature of the public Cloud, may be under constant change. When home healthcare is entirely within the hospital's scope of management and control, the threats are well known and quantifiable. When it is placed within a public Cloud, the application becomes vulnerable to not only the threats of the hospital but also the threats faced by the public Cloud on a global scale. Furthermore, the healthcare application can be threatened by non-technical threats such as destabilized foreign governments, foreign economy crashes, unforeseen natural disasters, wars, etc. It is therefore contended that the threat profile of placing home healthcare into the public Cloud could be negatively impacted.

• *Organizational ISMS and risk analysis*. As part of the hospital's ISMS, knowledge of all assets, asset management, and deployed controls and countermeasures is a required factor contributing to organizational risk registers. The risk registers are a statutory requirement as part of critical national infrastructure and are essential to influence future IT decisions and security plans. It is thought unlikely that a Cloud provider would divulge to its client the exact nature of its security controls, for to do so would illustrate a provider's capability and this information placed into the wrong hands could prove useful to a potential attacker.

• *Multi-tenancy challenges*. There is a concern that significant security threats can occur where the healthcare assets are held upon shared resources within the Cloud infrastructure. Multi-tenancy can open up the application assets to potential threats, such as:

– Potential loss of technical data separation controls leading to data leakage.

– Either a lowest common denominator or highest common denominator security control suite will apply to protect all tenant assets potentially leaving the healthcare assets under- or overprotected.

– Encryption can be used to enforce confidentiality and integrity within this model but therein lie key management problems. How best can cryptographic keys be managed within the public Cloud?

– How to specifically stipulate who you will absolutely not want to share resources with (for example a lighting industry competitor).

• *Intrinsic and extrinsic assurance*. One of the main challenges when outsourcing the SLA to a public Cloud is to gain assurance that the public Cloud provider will maintain the healthcare application in a manner contractually agreed. There are a number of mechanisms by which assurance can be gleaned, as follows:

– It is possible to gain a degree of extrinsic assurance of the technical controls deployed with the healthcare application upon a public Cloud if technical devices used are subject to formal evaluation and the architectural details (such as low-level designs) can be shared with the hospital. The use of a trusted computing infrastructure (TCI) can yield a degree of assurance. Furthermore, certification to international standards such as the ISO2700110 can go a long way towards assuring the credibility of the governance procedures and technical controls in place within a public Cloud provider.

– It is far more challenging to gain intrinsic assurance whereby one would wish to employ a 'right of inspection' upon any site with access to or processing the healthcare assets. The details of this could be to (either as an organization or via an approved third party) physically inspect sites, conduct penetration testing (to detect insecure config-uration, hard/software, APIs, etc.) and compliance audits with financial penalties for non-compliance. If chosen, these mechanisms would need to be placed explicitly into a legally binding framework.

• *Contractual issues.* Of all the challenges within this question, one is a lynchpin to ensure that security processes are followed correctly. For all security policies, procedures, and controls to be mandated, they must be placed into a legally binding contract that will apply throughout the legal jurisdictions where each party involved in the outsourcing of the application conducts business. This may mean that multiple contracts must be produced, containing the same pertinent information, for each legal jurisdiction. Furthermore, com-pliance documentation – such as codes of connection, memoranda of understanding, forms of understanding, and security operating procedures – must be bound to each contract to facilitate enforcement.

• *Business tie-in.* Owing to the federation of the public Cloud and its partnership, there is a challenge to overcome to ensure that the application developed upon the public Cloud can be moved to other Cloud providers as and when business requirements dictate (e.g., as in the case of unacceptable increase of Cloud charges, the public Cloud going out of business, etc.). Should the home healthcare application prove difficult or impossible to migrate to another provider for technical development constraints, the hospital could find itself tied into an unworkable contract with no obvious way to withdraw without significant economic expenditure.

References

[1] Thomas Ristenpart, Eran Tromer, Hovav Shacham, and Stefan Savage. Hey, you, get off of my cloud: Exploring information leakage in third-party compute clouds. In *Proceedings of the 16th ACM Conference on Computer and Communications Security, CCS '09*, pp. 199–212. ACM: New York, 2009.

[2] Yung-Chuan Lee, Stephen Bishop, Hamed Okhravi, and Shahram Rahimi. Information leakage detection in distributed systems using software agents. In *Proceedings of the International Conference on Intelligent Agents*, pp. 128–135. IEEE, 2009.

[3] Hamed Okhravi, Stephen Bishop, Shahram Rahimi, and Yung-Chuan Lee. A MA-based system for information leakage detection in distributed systems. In *Emerging Technologies, Robotics and Control Systems*, 3rd edn. 2009.

Index

Cloud Management and Security, First Edition. Imad M. Abbadi.
© 2014 John Wiley & Sons, Ltd. Published 2014 by John Wiley & Sons, Ltd.
Companion Website: www.wiley.com/go/abbadi_cloud

Printed and bound by CPI Group (UK) Ltd, Croydon, CR0 4YY

12/01/2025

14624501-0001